Advances in Horticulture Science Research

Advances in Horticulture Science Research

R.C. Upadhyaya

ANMOL PUBLICATIONS PVT. LTD.
NEW DELHI-110 002 (INDIA)

ANMOL PUBLICATIONS PVT. LTD.

H.O.: 4374/4B, Ansari Road, Darya Ganj,
New Delhi-110 002 (India)
Ph.: 23278000, 23261597

B.O.: No. 1015, Ist Main Road, BSK IIIrd Stage
IIIrd Phase, IIIrd Block
Bangalore - 560 085 (India)
Visit us at: www.anmolpublications.com

Advances in Horticulture Science Research

First Published, 2008

ISBN 978-81-261-3430-4

PRINTED IN INDIA

Printed at Mehra Offset Press, Delhi.

Contents

Preface

Horticulture has a special significance for the economic development. Most of the people, though hard working and face the calamities of nature and live in harsh conditions, yet produce lot of fruits, vegetables and other flowering and ornamental plants including the medicinal and aromatic plants. In order to increase the productivity of horticultural crops growers must take the advantage of the expertise of scientists. Keeping this idea in focus, the present work has been compiled which contains experiences of the experts and related people. The volumes cover advances in strategies, production, plant protection, value addition and other important areas of horticulture development.

The Encyclopaedia provides practical information on production technologies of tropical and subtropical fruits as well as plantation crops and horticultural nursery. Post-harvest management, marketing and export trends have been furnished for guidance. It also presents most realistic cost estimates and returns from cultivation of such crops. The volumes aim at viewing commercial prospects for horticulture. It has been carefully designed to meet the queries of intending investors, financial institutes, creditors, traders and growers engaged in dealing with horticultural crops. Further it will serve the purpose reference books to students, academicians and related people.

R.C. Upadhyaya

1

Horti Vision 2020

Assam is a traditionally horticultural State and the socio–economic fabric of its people is largely constituted by horticultural components. The horticultural beauty lies in its history, culture, traditions and in the mindset of the people–which has imbibed a lot from this colourful science of fruits, flowers, vegetables, spices, plantation crops, root & tuber crops, nut crops, medicinals, aromatics, mushroom, bee keeping and lately bamboo. With a mere acrage of 5.46 lakh hectors under horticulture, out of gross cropped area of 36.37 lakh ha, it is only 15% of this area but has a lot to say if we are to deliberate in terms of its economic and social contribution and more importantly, to see Horticulture in accordance with aspirations and expectations of coming generation in the years to come or precisely by the year 2020 and beyond.

Horticulture has a commitment that no other sector may perhaps can come up as parallel i.e. in terms of its importance to achieve nutritional security and as a most viable component of environment friendly industrial base.

This is a sector gaining importance day by day and the reason behind for calling it as sunrise area are – diverse nature of its products to suit different agro climatic situation, higher bio mass production per unit area and thereby more income per unit area, scope for improvement in rural diet with least investment or any extra effort, scope for massive employment generation, unveiling new horizons of value addition and the science itself being intellectually satisfying with aesthetic outlook has an universal appeal for both developed and developing nations. The situation can be looked under Assam's perspective as here under:

1. Horticultural sector which includes fruits, vegetables, tuber crops, floriculture, mushroom, medicinal & aromatic plants, spices and plantation crops have proved beyond doubt to be the best diversification of agriculture for better land use.
2. The soil and climate is so ideal for a wide range of horticultural crops that experts comment Assam as a 'Sleeping Giant'.
3. According to latest survey on children aged 1 to 4 and pregnant women in Assam, 50.4% of them suffer from under weight and the growth of 52.5% are stunted. So individual dietary improvement through nutritional horticulture is the need of the hour-which is easily achievable and most viable.
4. Tremendous scope for employment generation including farm-women through horticultural ventures & self employment horticultural schemes.
5. Sustainable approach to raise productivity is best suited when horticulture is taken into account due our existing production system approach.
6. The natural flora of Assam have huge number of hitherto unexploited medicinal and aromatic plants including minor fruits and orchids. Any strategy to boost up this sector will particularly help local and more particularly tribal farmers. This has a bearing on preserving genetic variability that exists in case of many horticultural crops also.
7. The State being characterized for low consumption of fertilizer and plant protection chemicals, this so called weakness can be transformed in to strength through adoption of organic farming as the export tract is fast becoming clear for organics only.
8. Assam being gateway of North East India and the State being strategically located in close proximity to emerging South East Asian markets, has a lot to deliver in years to come. Keeping these points in view and also considering the requirement against projected population of Assam by the year 2020 and giving top most priority on

commercialization of this sectors, crops specific strategies are formulated and a gist of targets of various major sectors of horticulture has been formulated as stated below.

Banana

Highest area under fruit crops is occupied by banana. Out of more than 15 varieties grown, the commercially important varieties are Jahaji, Malbhog, Amrit sagar, Kach kol and Cheni cahampa. Marketable surplus is high and sizeable quantity is going outside the state. Commercial banana activity is common in and around Drangiri of Goalpara district. Jorhat district is known for abundance of best quality, Jahaji while Nagaon, Morigaon, Kamrup districts are referred as producers of best quality Malbhog. There is scope for more organised cultivation and better market tie-up to ensure price support to the growers.

1. Increase in area from present 42.6 thousand Ha to potential 70.0 thousand Ha by the year 2020 with an annual growth rate of 3.78%
2. Thrust districts identified are Goalpara, Kamrup, Nagaon, Morigaon and Jorhat.
3. Hike in productivity and commercialization to be achieved by:
   i. Identification of potential belts & organization of banana growers to from Growers societies and their training/awareness/exposure.
   ii. Development of commercial plantation with commercial varieties
   iii Adoption of package of practices
   iv. Adoption of modern technologies like Drip/Fertigation.
   v. Agronomic manipulation/mechanization
   vi. Adoption of IPNM technologies.
   vii. Reduction in post harvest losses with appropriate technologies including proper harvest, grading, packaging, storage.
   viii. Facilitate marketing/market information/market tie-up/for domestic and export market.

4. Quality improvement.
5. Value addition.

Local cultivars are of excellent quality but it is grown mostly as homestead and commercial cultivation is only in small scale. It is highly remunerative during Off-season i.e. February, March, April and after reaching a plateau, the price becomes unremunerative. Almost continuation bearing habit, least demand for space, suitability as intercrop makes it very potential crop for augmenting farmers income per unit area. Potential districts are Kamrup, Sonitpur, Barpeta Bongaigaon, Nagaon. Dhubri, Goalpara, Kokrajhar. There is scope for organised cultivation for processing products to ensure price support to the growers.

Can be grown as boarder crop for protection of orchard.

Strategies for increasing production and quality:

1. Increase in area from present 8.80 thousand Ha to potential 17.50 thousand Ha by 2020 with an annual growth rate of 5.81%.
2. Thrust districts identified are Kamrup, Barpeta, Bongaigaon, Sonitpur
3. Assam Lemon will be a major component for multi storied cropping.
4. Parallee thrust is semi urban and urban areas.
5. Hike in productivity and commercialization to be achieved by:-
   - i. Identification of potential belts & training/awareness to the growers.
   - ii. Development of commercial plantation with suitable varieties.
   - iii. Adoption of package of practices.
   - iv. Intercropping with other horticultural crop arecanut, coconut etc.
   - v. Post harvest processing for pickles, squash with proper packaging, storage.
   - vi. Facilitate marketing/market information/market tie-up/for domestic and export market.

Orange

Broad Objective:

(a) Self–sufficiency and market exploration and commercialization.

(b) Better price to growers.

(c) Value addition.

Local cultivars are of excellent quality but area is gradually declining due to lack of market support, maladies associated with "citrus decline" and above all, encroachment of traditional orange area by Mini Tea Garden since last one decade for obvious reasons of lucrative prevailing tea prices while prices of orange is totally unattractive, unstable/unpredictable governed by middleman.

Strategies for increasing production and quality:

1. Increase in area from present: 5.96 thousand ha to potential 10 thousand ha by the year 2020 with on annual growth rate of 3.98%
2. Rejuvenation of existing orchards
3. Thrust districts identified are Tinsukia, N.C. Hills, Karbi anglong, Darrang, Kamrup, Goalpara.
4. Hike in productivity and commercialization to be achieved by:-
   - i. Identification of potential belts & organization of Orange growers to farm Growers societies and their training/awareness/exposure.
   - ii. Development of commercial plantation & rejuvenation of old orchards.
   - iii. Adoption of package of practices
   - iv. Adoption of IPNM technologies.
   - v. Organic cultivation of orange for export market
   - vi. Reduction in post harvest losses with appropriate technologies including proper harvest grading, packaging, storage.
   - vii. Facilitate marketing/market information/market tie-up/for domestic and export market.

5. Quality improvement.
6. Value addition.

Litchi

Broad Objective:

(a) Entering into competitive domestic market with quality product.

(b) Development of commercial plantation

(c) Better price to growers.

(d) Value addition.

Wide range of varieties are grown and most of them are excellent in quality; very much identical to Muzaffurpur. Qualitatively, the varieties grown in Sonitpur district may surpass any elite variety. The crop being highly perishable, post harvest losses are common and fail to reach city market as garden fresh and thus price normally becomes unremunerative. Strategies for increasing production and quality:

1. Increase in area from present 4.12 thousand ha to potential 6.77 thousand ha by the year 2020 with an annual growth rate of 3.78%.
2. Thrust districts identified are Sonitpur, Kamrup, Bongaigaon, Jorhat and Goalpara
3. Hike in productivity and commercialization to be achieved by:
   - i. Identification of potential belts & organization of Litchi growers and their training/awareness/exposure.
   - ii. Development of commercial plantation with commercial varieties
   - iii. Adoption of package of practices
   - iv. Adoption of IPNM technologies.
   - v. Reduction in post harvest losses with appropriate technologies to protect from bird and bats with plastic nests.
   - vi. Facilitate marketing/market information/market tie-up/for domestic and export market.
4. Value addition.

Mango

Broad objective:

(a) Development of commercial plantation with elite varieties.

(b) Better price to growers.

(c) Value addition.

Inspite of being an major fruit crop, it is not commercially cultivated. Cultivation is confined to few districts in homestead garden only with local cultivars which is not competitive with the imported (other state) one.

At present few villages have been selected as Mango village with supply and introduction of improved varieties like Amrapalli, Doseri, Mallika etc. Local varieties being infested by fruit borers, utilized mostly for pickle production

Strategies for increasing production and quality:

1. Increase in area from present 3.70 thousand ha to potential 10.95 thousand ha by the year 2020 with an annual growth rate of 11.52 %
2. Thrust districts identified are Kamrup, Sonitpur, Bongaigaon, Goalpara, Kokrajhar, Golaghat and Nagaon.
3. Hike in productivity and commercialization to be achieved by:-
    i. Identification of potential belts & villages and organization of Mango growers to form Growers societies and their training/awareness/exposure.
    ii. Development of commercial plantation with commercial varieties
    iii. Adoption of package of practices
    iv. Adoption of IPNM technologies.
    v. Reduction in post harvest losses with appropriate technologies.
    vi. Facilitate marketing/market information/market tie-up/for domestic and export market.
4. Quality improvement.
5. Value addition.

Papaya

Broad objective:

(a) Development of commercial Plantation.

(b) Better price to growers.

(c) Value addition.

In Assam, Papaya is a traditional homestead crop and commercialization is not yet started.

Varieties grown are almost local and improved hybrid varieties grown only in small scale. The present use of Papaya is mostly as vegetables and not popular as industrial crops.

Strategies for increasing production and quality:

1. Increase in area from present 6.94 thousand Ha to potential 14.00 thousand Ha by the year 2020 with an annual growth rate of 5.98%.
2. Thrust districts identified are: N. C. Hill, Kamrup, Nalbari, Kokrajhar, Darrang, Cachar, Karimganj, Nagaon, and Jorhat.
3. Hike in productivity and commercialization to be achieved by:-
    i. Identification of potential belts & villages and organization of training/awareness/exposure.
    ii. Development of commercial plantation with commercial varieties
    iii. Adoption of package of practices
    iv. Adoption of IPNM technologies.
    v. Reduction in post harvest losses with appropriate technologies including proper harvest, packaging.
    vi. Exploring Industrial uses primarily through small scale entrepreneurs
    vii. Facilitate marketing/market information/market tie-up and exploring contractual farming.
4. Quality improvement.
5. Value addition.

Pineapple

Broad Objective:

(a) Self–sufficiency and market exploration and commercialization.

(b) Better price to growers.

(c) Value addition.

Local cultivars grown are Kew (70%) queen and Joldhup. The quality is ideal for fresh as well as processed market and mostly organic. The crop is unremunerative and distress sale and spoilage during peak period is common. Weak processing sector is also a major problem. Strategies for increasing production and quality:

1. Increase in area from present 13.89 thousand ha to potential 22.50 thousand ha by the year 2020 with an annual growth rate of 3.64%
2. Thrust districts identified are Kamrup, Karbi Anglong, N.C. Hill, Nagaon, Goalpara and Cachar.
3. Hike in productivity and commercialization to be achieved by:-
   i. Identification of potential belts & organization of Pineapple growers to form Growrs societies and their training/awareness/exposure.
   ii. Development of commercial plantation with commercial varieties as monocrop/intercrop
   iii. Adoption of package of practices
   iv. Adoption of modern technologies like plastic mulching, Fertigation through drip etc.
   v. Agronomic manipulation for yield and extension of harvesting period to escape glut and supplement processing sector.
   vi. Adoption of IPNM technologies.
   vii. Reduction in post harvest losses with appropriate technologies.
   viii. Facilitate marketing/market information/market tie-up/for domestic and export market.
4. Quality improvement.
5. Value addition.

Other Fruits

Jalphai, Outenga, Kujithekera, Amara, Silikha, Jamun, Aonla, Bael, Castard Apple etc.)

Broad objective:

(a) Strengthening production base & market exploration and commercialization.

(b) Better price to growers.

(c) Exploring Value addition and MNC Tie-up.

In Assam many minor fruits are grown in homesteads,. Commercial cultivation is not started though these fruits have high nutrative and medicinal value and there is demand for raw consumption and for preparation of pickles, Jam Jally etc. The major fruits are:- Jalphai (Elaoocarpus florgundurs), Outanga (Dillenia indica), Silikha (Terminalia Chebela), Aonla (Emblica officinalis), Jamun (Syzygium cumini), Bael (Aegle marmelos), Castard apple (Annong squomosa) Etc.

Strategies for increasing production and quality:

1. Increase in area from present 1.95 thousand Ha to potential 3.00 thousand Ha by the year 2020 with an annual growth rate of 3.16%.
2. Thrust districts identified are almost all the district of the State.
3. Hike in productivity and commercialization to be achieved by:-
   i. Identification of potential belts, Villages & organization of farmers for commercial production.
   ii. Reduction in post harvest losses.
   iii. Facilitate marketing/market information/market tie-up/for domestic and export market.
4. Evaluation of local germplasm with research support.
5. Value addition.

Cashewnut

Broad objective:

(a) Large scale introduction in non traditional bu-potential areas in districts to strengthen production base.

(b) Marginal & waste land utilization.

(c) Development of commercial plantation to augment farmers income and develop processing unit.

Assam is not incorporated in the Cashewnut map of India but the fact remains that as many as 10 Nos. of cashew processing factories are running in Assam. The success of this crop is demonstrated in Mankachar area adjoining Garo hills of Meghalaya. Like tea, the processing sector being well organized, cashew crop may dominate other crops in most of the districts as even marginal lands can fetch an income of Rs. 1,00,000/- ha. when cashew is grown.

Strategies to boost up this sector:

i. Identification of potential belts particularly in areas adjoining Arunachal Pradesh, Meghalaya etc. and hill districts and gist fallow lands.

ii. Strengthening of nursery sector with elite planting material.

iii. Explore mix cropping/inter cropping to accommodate cashewnut with parallel thrust in mono cropping to strengthening production base.

iv. Processing under private/cooperation sector.

v. Facilitate marketing/market tie-up.

vi. Quality improvement.

Coconut

Broad objective:

(a) Strengthening production base by encompassing non traditional but potential areas.

(b) Better price to growers.

(c) Explore value addition.

Though Assam is suitable for Coconut cultivation but the productivity is low as non descriptive local varieties are mostly grown due to non availability of good quality Coconut seedlings within easy reach of farmers, very poor management practices in plantations, affect of diseases more particularly by Ganodema, Borax deficiency etc. Moreover, the coconut farmers are also not getting attractive prices due to non existence of coconut based industries and proper market tie up. Ongoing scheme of Coconut Development Board is providing impetus to this sector. Performance is excellent in districts of Kamrup, Nagaon, Morigaon, Nalbari and Barpeta.

Strategies for increasing production and quality:

1. Establishment of Coconut Nurseries for availability of best planting materials for the farmers.
2. Increase in area from present 21367 thousand Ha to potential 26967 thousand Ha through area expansion inter cropping, mixed cropping etc. by the year 2019-20.
3. Increase in productivity from present 73 to 95 Nuts per bearing tree through introduction of improved technologies.
4. To Identify the potential coconut pockets followed by thrust programme for higher productive and to establish coconut based industries like tender coconut water, coir etc.
5. Management of diseased/unproductive/old coconut garden and need base replacement with new seedling
6. Introduction of multitier and farming in coconut garden for attractive benefit to the farmers.
7. Establishment of net work for market tie-up and promotional activities.

Floriculture

Broad Objective:

(a) To check flow from neighbouring States

(b) To fulfil ever increasing domestic market demand.

(c) To explore external/export market

(d) Complete commercialization to augment farmers income.

Floriculture is a sunrise area in Assam. The area is fast expanding, commercial production has started. Major components at present are Marigold, Gladioli, Tuberose, Gerbera and recently Anthurium and foiages in the form of fresh flower, garland, cut flowers, dry flower, Buquecy, potted pants. Farmers more particularly farm woman of Kamrup district are entering into commercial floriculture followed by Morigaon, Nagaon and Tinsukia & Cachar. The major problems are lack of off season production technology and very poor post harvest handling practices. If these two problem are properly addressed, and if local orchids & ferns can be explored, floriculture has not only a very bright prospect in Assam but can explore foreign markets.

Strategies for increasing production and quality:

1. Advantage Guwahati *i.e.* exploring most prospective market of Guwahati due to its strategic location for floriculture products will be overall target.
2. Identification of Floriculture potential villages more particularly in Kamrup & districts surrounding Kamrup, formation of growers societies, their training and exposure
3. Saturation of floriculture area with better crops/Specie/ varieties having market demand.
4. Explore local germplasm particularly orchids/ferns etc. with research support.
5. Easy availability of planting materials.
6. Adoption of improved and hi-tech technologies for productivity hike and off-season production.
7. Quality improvement

Exploring Market/Market information/Market tie up.

Rabi Vegetables

Broad objective:

(a) Self –sufficiency and market exploration and commercialization.

(b) Better price to growers.

(c) Restricted use of Chemicals

(d) Value addition.

Rabi vegetables in Assam have already shifted from tiny homestead to large scale commercial plantation in almost all districts. Hybrids being admired and cultivated much, market glut at peak season has become a problem in case of vegetable crops. High perishability, lack of storage facility, poor communication has contributed to the malady. Besides common

Cole crops, solanacius vegetable, new crops like capsicum, Broccoli and new varieties are dominating the production and market scenario.

Strategies for increasing production and quality:

1. Increase in area from present 142.62 thousand Ha to potential 356.80 thousand Ha by the year 2020 with an annual growth rate of 8.83%
2. Thrust districts identified are: Barpeta, Dhubri, Bongaigoan, Darrang, Kamrup, Goalpara, Nagaon, Morigaon & Cachar.
3. Hike in productivity and commercialization to be achieved by:-
   i. Identification of potential belts & organization of Vegetable growers to form Growers societies and their training
   ii. Off season production
   iii. Development of commercial plantation with commercial varieties
   iv. Adoption of package of practices
   v. Agronomic manipulation/mechanization
   vi. Adoption of IPNM technologies and organic farming.
   vii. Reduction in post harvest losses with appropriate technologies including packaging, storage.
   viii. Facilitate marketing/market information/market tie-up/for domestic and export market.
4. Introduction of new potential crops
5. Quality improvement.
6. Value addition and encouragement of contractual farming.

Potato

Broad objective:

(a) Self–sufficiency and market exploration and commercialization.

(b) Better price to growers.

(c) Value addition.

Though Potato in Assam have already been started on a commercial scale in many districts, the present availability is below our requirement. But market glut at peak season has become a problem in case of potato. Perishability lack of storage contributed to less return to the growers.

Strategies for increasing production and quality:

1. Increase in area from present 75.48 thousand Ha to potential 140.23 thousand Ha by the year 2020 with an annual growth rate of 5.04%.
2. Thrust districts identified are: Goalpara, Kamrup, Barpeta, Morigaon, Dhubri, Darrang and Sonitpur.
3. Hike in productivity and commercialization to be achieved by:-
   - i. Identification of potential belts & organization of Potato growers to form Growers societies and their training/awareness/exposure.
   - ii. Development of commercial cultivation with commercial varieties.
   - iii. Adoption of package of practices
   - iv. Agronomic manipulation/mechanization
   - v. Adoption of IPNM technologies.
   - vi. Reduction in post harvest losses with appropriate technologies including proper harvest, grading, packaging, storage.
   - vii. Facilitate marketing/market information/market tie-up/for domestic and export market.
4. Easy availability of standard planting material.
5. Quality improvement
6. Value addition.
7. Emphasis on TPS technology.

Kharif Vegetables

Broad objective:

(a) Self–sufficiency and market exploration and commercialization.

(b) Better price to growers.

(c) Value addition.

Kharif vegetables in Assam is a profitable venture now more particularly the Off season production. Ladies finger, spine gourd, snake gourd, smooth gourd and other cucurbits are most sought after crop. Hybrids being admired much, market glut at peak season has become a problem in case of Kharif vegetable crops. High perishability, lack of storage facility, poor communication has contributed to the malady. However, district like Barpeta, Darrang, Kamrup are far ahead of other districts in case of commercial vegetable cultivation.

Strategies for increasing production and quality:

1. Increase in area from present 67.31 thousand Ha to potential 184.00 thousand Ha with an annual growth rate of 10.19%
2. Thrust districts identified are: Barpeta, Dhubri, Darrang, Kamrup, Nalbari, Goalpara, Nagaon, Morigaon.
3. Hike in productivity and commercialization to be achieved by:-
    i. Identification of potential belts & organization of Vegetable growers to form Growers societies and their training
    ii. Development of commercial plantation with commercial varieties
    iii. Adoption of package of practices
    iv. Agronomic manipulation/mechanization
    v. Adoption of IPNM technologies.
    vi. Reduction in post harvest losses with appropriate technologies including packaging, storage.
    vii. Facilitate marketing/market information/market tie-up/for domestic and export market.
    viii. Minimize the involvement of middleman through the organization of vegetable growers.
4. Protected cultivation
5. Quality improvement.
6. Value added.

Black Paper

Broad objective:

i. Entering into domestic and external market.

ii. Better price to growers.

iii. Processing and Value addition.

Black Pepper is a common component in homestead in Assam and seldom finds a commercial status. In spite of assured market and its non-perishable nature, easy growing habit in arecanut and coconut lantations, this commercial giant could not draw the attention of common Assamese farmers. A simple estimate shows that per ha. return is more than Rs. 100,000/-.

Strategies for increasing production and quality:

1. Increase in area from present 2.70 thousand Ha to potential 6.80 thousand Ha by the year 2020 with an annual growth rate of 8.93%
2. Thrust districts identified are: All the district are more or less equally potential but the location should be free form flood & water stagnation
3. Hike in productivity and commercialization to be achieved by:-
   i. Identification of potential belts & villages on the basis of available standards like Arecanut, Coconut etc. and organization of Black pepper growers societies.
   ii. Development of commercial plantation with commercial varieties.
   iii. Adoption of package of practices.
   iv. Agronomic manipulation.
   v. Adoption of IPNM technologies.
   vi. Reduction in post harvest losses with appropriate technologies including proper harvest grading, packaging, storage.
   vii. Facilitate marketing/market information/market tie-up/for domestic and export market.
4. Quality improvement.
5. Value addition like white pepper, oil, oleoresins etc.

Chilly

Broad objective:

(a) Commercialization.

(b) Better price to growers.

(c) Value addition.

It is common spice crop in Assam which is widely grown specially in Rabi season in Nagaon, Morigaon, Darrang, Sonitpur, Nalbari, Barpeta, Kokrajhar, Goalpara and Dhubri district.

It has a local market which is used as raw and dry. But there is a scope for replacement with improve varieties to meet the demand. There is scope for more organized cultivation and better market tie up to ensure price support to the growers. One local capsicum species locally know as Bhut Jalakia is reported to be the most pungent spices and the ighest capsis content projecting scope for commercialization. Another species, Birds eye chilli is grown abundantly in Karbi Anglong, N.C. Hills and Barak valley and exported to Bangladesh.

Strategies for increasing production and quality:

1. Increase in area from present 14.88 thousand Ha to potential 27.0 thousand Ha by the year 2020 with an annual growth rate of 4.79%.
2. Thrust districts identified are Nagaon, Morigaon, Barpeta, Dhubri, Cachar and Bongaigaon.
3. Hike in productivity and commercialization to be achieved by:-
    i. Identification of potential belts & organization of Chilli growers and their training/awareness/exposure.
    ii. Adoption of package of practices
    iii. Adoption of IPNM technologies.
    iv. Facilitate marketing/market information/market tie-up/for domestic and export market.
4. Quality improvement.
5. Value addition.

Corriender

Broad objective:

i. Large scale commercial adoption in revering tracts

ii. Better price to growers.

iii. Processing.

Coriander is a neglected but potential crop in Assam. Small scale house hold production is common in Rabi season but the riverain tract produces substantial quantity annually as seeds but not processed locally. Competition with other field spices like cumin, flannel, fenugreek etc. restricting its expansion.

Strategies for increasing production and quality:

1. Increase in area from present 19.90 thousand Ha to potential 31.50 thousand Ha by the year 2020 with an annual growth rate of 3.42%
2. Thrust districts identified are: Barpeta, Nalbari, Morigaon, Dhubri, Darrang, Goalpara, nagaon, Bongaigaon and all the districts have potentiality to cultivate corriandar.
3. Hike in productivity and commercialization to be achieved by:-
   i. Identification of potential belts & organization of growers and their training/awareness/exposure.
   ii. Development of commercial cultivation with commercial varieties.
   iii. Adoption of package of practices.
   iv. Facilitate marketing/market information/market tie-up/for domestic and export market.
4. Value addition.

Ginger

Broad objective:

(a) Self–sufficiency and market exploration and commercialization.

(b) Better price to growers.

(c) Value addition.

Inspike of being a major spice crop commercial cultivation is mostly confined to two hill district and in Sadia sub-division of Tinsukia district. Local varieties dominated in most districts. Moran Ada is admired for high leoresin content; while Rio-de-generio dominates in Sadia area.

Mizo lange variety is predominantly grown in Karbi anglong and N.C. Hills. Both these varieties have export potential and also well fitted for processing and value addition. The fore most obstacle in commercialization of this sector is poor communication to the centres of production and as such paving the way for middleman exploitation and lack of processing units.

Strategies for increasing production and quality:

1. Increase in area from present 19.97 thousand Ha to potential 35.70 thousand Ha with an annual growth rate of 5.80%
2. Thrust districts identified are: Karbi Anglong, Tinsukia, N.C. Hills, Darrang, Kamrup, Goalpara, Nagaon, Cachar.
3. Hike in productivity and commercialization to be achieved by:-
   i. Identification of potential belts & organization of Ginger growers to from Growers societies and their training/awareness
   ii. Development of commercial plantation with commercial varieties
   iii. Adoption of package of practices
   iv. Agronomic manipulation/mechanization
   v. Adoption of IPNM technologies.
   vi. Reduction in post harvest losses with appropriate technologies including proper harvest, grading, packaging, storage & processing.
   vii. Facilitate marketing/market information/market tie-up/for domestic and export market.
4. Quality improvement.
5. Value addition like Ginger sticks, pastes, oil, oleoresins etc.

Garlic

Broad objective:

(a) Development of commercial plantation in potential pocket.

(b) Better price to growers.

(c) Value addition.

Though the garlic is a suitable crop for Nagaon, Morigaon, Jorhat, Kamrup, Darrang, Sonitpur, Goalpara, Bongaigaon and Dhubri district it is not cultivated widely. The local cultivars are dominating the area. The crop is yet to be cultivated commercially.

Strategies for increasing production and quality:

1. Increase in area from present 6.69 thousand Ha to potential 12.40 thousand Ha by the year 2020 with an annual growth rate of 5.02%
2. Thrust districts identified are Goalpara, Jorhat (Majuli area), Barpeta, Dhubri, Bongaigaon, Goalpara, Darrang and Kamrup.
3. Hike in productivity and commercialization to be achieved by:-
   i. Identification of potential belts & villages and organization of training/awareness/exposure for farmers.
   ii. Development of commercial plantation with commercial varieties.
   iii. Adoption of package of practices
   iv. Facilitate marketing.
4. Quality improvement.
5. Value addition.

Onion

Broad objective:

(a) Self–sufficiency and market exploration and commercialization.

(b) Better price to growers.

(c) Value addition.

Onion is a back yard cultivation in Assam & could hardly meet even the seasonal requirement. Rabi Onion is the only crop while Kharif onion, as per preliminery studies, is promising but in sloppy lands and foot hills

Strategies for increasing production and quality:

1. Increase in area from present 7.50 thousand Ha to potential 12.70 thousand Ha with an annual growth rate of 4.07%
2. Thrust districts identified are: Barpeta, Bongaigaon, Darrang, Dhubri, Kamrup, Nagaon.
3. Hike in productivity and commercialization to be achieved by:-
   - i. Identification of potential belts and villages & organization of farmers for large scale cultivation and their training, awareness/exposure
   - ii. Introduction of commercial varieties
   - iii. Adoption of package of practices
   - iv. Adoption of IPNM technologies.
   - v. Facilitate marketing
4. Quality improvement.

Other Spices

Broad objective

(a) Self–sufficiency and market exploration and commercialization.

(b) Introduction of better varieties.

(c) Better price to growers.

(d) Value addition.

In Assam there are some areas under other spices like Cumin, Black Cumin, Flannel, Fenugreek etc. but it has not yet been commercialized in spite of great demand and lucrative market. The area is mostly confined to riverain tract and product is exclusively handled by middleman. The post harvest aspects are neglected and quality of product needs improvement.

Strategies for increasing production and quality:

1. Increase in area from present 2.12 thousand Ha to potential 4.70 thousand Ha by the year **2020** with an annual growth rate of 7.15%.
2. Thrust districts identified are Goalpara, Kamrup, Nagaon, Morigaon, Darrang, Sonitpur, Dhubri and Barpeta.
3. Hike in productivity and commercialization to be achieved by:
   i. Identification of potential belts and their training/awareness/exposure.
   ii. Development of commercial plantation with commercial varieties
   iii. Adoption of package of practices
   iv. Reduction in post harvest losses with appropriate technologies.
   v. Facilitate marketing/market information/market tie-up/for domestic and export market.
4. Quality improvement.
5. Value addition.

Turmeric

Broad objective:

(a) Commercialization.

(b) Better price to growers.

(c) Value addition.

Though Turmeric is a house hold crop in Assam, commercial cultivation has started only in small scale. The present productivity is very low and whatever produced is utilized for mere local consumption.

Strategies for increasing production and quality:

1. Increase in area from present 14.88 thousand Ha to potential 27.00 thousand Ha by the year 2020 with an annual growth rate of 4.79%
2. Thrust districts identified are: Karbi Anglong, N.C. Hill, Kamrup, Golaghat, Nagaon, Morigaon, Lakhimpur, Sonitpur, Kokrajhar.

3. Hike in productivity and commercialization to be achieved by:
   i. Large-scale establishment of commercial plantation also as intercrop.
   ii. Saturation with commercial varieties.
   iii. Adoption of package of practices
   iv. Intercropping with Banana, Arecanut, Coconut.
   v. Adoption of IPNM technologies and organic production.
   vi. Adoption of appropriate technologies for post harvest processing, packaging, storage.
   vii. Facilitate marketing/market information/market tie-up/for domestic and export market.
4. Quality improvement.
5. Value addition.

Medicinal and Aromatic Plants

Broad objective:

(a) Evaluation of local germ plasm and introduction of potential exotics

(b) Prioritization on commercial outlook.

(c) Systematic cultivation/contractual farming/forest village exploration.

(d) Processing of final or semi processed product

(e) Exploring markets/market information/market tie-up.

Assam being one of the 14 bio diversity hot spots of the world, it is the natural depository of many valuable medicinal and aromatic plant partially explored or still to be explored commercially. As reported, many valuable items are going out of the state informally by traders. There is no systematic production of marketing or any processing unit on MAP. Recently, contractual farming is being encouraged by private sector,. Research support wanting.

Strategies to boost up this sector:

1. Prioritization of commercially important crops/varieties/ Species with research support.

2. Organized cultivation with systematic approach
3. Exploring Joint Forest Management Committee and others to venture into commercial production/processing/ semi processed product manufacture/value addition etc.
4. Market access/market information
5. Linking private sector and contractual farming.

Bambo

Time has come to rediscover bamboo. It is no longer a "orphan specis" or poor men's timber. Pondits in World Bamboo Congress (27.02.2004 to 04.03.04 at New Delhi) have projected this crop as 'green gold' and referred it as vital for social, ecological and economical security. No other crop has more than 1500 application such as – Bamboo pulp & paper, laser printer and photocopy paper, floor boards, panels, ceiling, medicinal, receipies, cloths, land escaping, eco tourism, building, furniture etc. The annual turnover on bamboo sector projected as Rs. 25 billion, against the market potential of Rs. 45 billion and it is expected to go beyond Rs. 250 billion in near future. India posses 140 specis *i.e.* just next to China but the productivity is less. Bamboo is reported to be a gold mine for gene tagging. Estimate says that out of 9 million Ha, 1 million ha of bamboo is in Assam. It is revealed that 66% of the total area under Bamboo in India is in the North East and Mizoram share is 33%. The present world market is dominated by china with its products like Flavonoid of Bamboo leaves, Bamboo juice, panel, veneer container bamboo fabric-proof ceiling (fire proof), Bamboo charcoal, vinegar, soil fungicide, cosmetics, soaps, handicrafts, musical instruments, bonsai, Bamboo garden etc.

It is suggested that there should be primary processing in N.E. India prefarbly Assam, the gateway of Northeast while the final product industries may be anywhere else. Against this back drop, time has come to outline modelities to develop bamboo sector in Assam. Bamboo isbeing declared as horticultural crop, the department has to adopt strategies and it is proposed as below:

1. Package of practice for bamboo and its economics is to be finalized and circulated targeting homestead areas.

2. Massive training/awareness Programme
3. Exposure of farmers/entrepreneurs on recent value addition of bamboo for development of bamboo based industries.
4. Nursery development and Area expansion.

Mushroom

Broad objective:

a. Developing mushroom as a major component for self employment, dietary and augment farmer's income.
b. Explore market demand for oyster & Button mushroom.
c. Value addition.

In Assam mushroom is known to many but there is very limited commercial production and lacking systematic approach. The Departmental mushroom spawn production laboratory now being run by private entrepreneurs could develop mushroom owner network under buy back system and the unit is gradually picking up. At present the annual production of fresh mushroom is 2000 tones only against a projected demand of 12,000 tones besides export potential. The total no. of mushroom entrepreneur in the state is about 60 nos but they are mostly seasonal. The most common edible mushroom which is in practice is Oyster mushroom.

Strategy for increasing production & Quality:

(i) Involving more farmers/SHG/Women SHG to venture in the commercial mushroom production.
(ii) Easy availability of spawn.
(iii) Satellites farming
(iv) Processing and value addition.

Processing, Value Addition and Marketing of Horticultural Products of Assam

Assam will have to be more fast in this sector in years to come. The crops that can be immediately targeted are various spices like Ginger, Black pepper, Field Spices, Fruits like Banana, Pineapple, Satkara (Cmarcroptera), Citrus crops, Vegetables, Medicinals and Aromatics, Bamboo etc. Before that it must address vital problems of volume and quality of its produce.

The strategy to boost up this sector will be:

(a) Strengthen production base
(b) Improvement in quality of produce
(c) Maximum possible extension of harvesting period
(d) Cultivation of product specific varieties
(e) Setting up of multi-product processing units – incorporating – Spice products, fruits, vegetable and even medicinal & aromatics.
(f) Setting up of packaging unit and exploring bio packaging & bio preservatives.
(g) Exploring possibility for setting up large or medium scale processing unit with tie-up with satellite processing units (Mini units) at interiors in pockets of production of raw materials.
(h) Adoption of cultivation & processing in accordance with domestic market demand and also visualizing emerging external markets, more particularly South East Asian markets through border trade due to phasing out of trade barriers amongst these countries and easy surface communication to be made as outcome of latest ASEAN & SAARC declarations.
(i) Entrepreneurship friendly rules, regulations policies and pattern of assistance.

Executive Summary

To achieve nutrition and income security for the people, particularly in arid region, suitable horticultural production technology is of vital importance. Although the Indian Council of Agricultural Research (ICAR) started multilocation collaborative research in 1976, major initiative on this aspect was taken in 1993 with the establishment of National Research Centre for Arid Horticulture (NRCAH now as CIAH) at Bikaner. This strategic plan for arid zone horticultural research updates the ongoing programme under the CIAH which is a part of the ICAR system under the Department of Agriculture Research and Education (DARE). "To conduct basic and applied research for improvement in productivity of horticultural crops and development of horticulture-based cropping system under arid environment; and to act as a repository of information related to arid horticulture, so as to achieve the following objectives:

To introduce, collect characterize, conserve and evaluate bio-diversity of horticultural crops under arid environment.

To utilize the available bio-diversity and improve the target crops to develop high quality and productive types having tolerance to biotic and abiotic stresses.

To study the factors related to rapid multiplication of propagules in case of established as well as new drops and the problems related to their growth and development.

To standardize agro-techniques with respect to efficient use of soil, water and nutrients for increased horticultural productivity.

To study the eco-physiological parameters of cropping system models for utilization of high temperature and radiation resources alongwith studies on mechanisms of physiological adaptation of horticultural crops to drought, heat and salt.

To develop post-harvest technology package for extended use of the horticultural produce of arid region.

To develop integrated pest and disease management technologies for horticultural crops under arid environment."

The programme will be built on the basis of multi-disciplinary projects comprising a series of activities. The developed technologies will be tested at relevant locations under the instrument of an All India Coordinated Research Project already in operation under the CIAH. Linkages to augment and strengthen execution of the programme will be maintained with several national and international organisations and developmental agencies.

The present provision of professional staff of 21 will need to be raised to a level of 60 during 1997-2002 to support the expanded activities. Accordingly, the budgetary requirements will rise to Rs. 1800 lakhs from the present level of Rs. 350 lakhs. Efforts will be made to generate internal resources to met part of the funding. Activities such as human resources development for research and development activities will also need to be strengthened.

Preamble

During the V plan and *ad hoc* Project under AP Cess Fund Scheme was started by the Indian Council of Agricultural Research (ICAR) in 1976 at ten centres for research on some selected fruits in the semi-arid areas in India. After relocation and addition of some centres during the VI Plan, this became a component of an ongoing All India Coordinated Fruit Improvement Project of the ICAR. With further strengthening and modifications during the VIII Plan, this component was carved out to form the All India Coordinated Research Project on Arid Zone Fruits (AICRP on AZF) to conduct multi-locational experiments on various aspect of fruits like *ber,* pomegranate, date palm, aonla, fig custard apple, bael, etc. at 12 centres in India, *viz.*, Abohar in Punjab, Aruppukottai in Tamil Nadu, Anantapur, in Andhra Pradesh, Bangalore in Karnataka, Bawal in Haryana, Rahuri in Maharashtra, Faizabad in Uttar Pradesh, Mundra and Sardarkrushinagar in Gujarat and Bikaner, Jodhpur and Jobner in Rajasthan. The work programme of this Project was thus confined to multi-locational testing of the technologies on some selected fruits being generated by the rather inadequate manpower and infrastructure inputs either in the ICAR Institutes such as Central Arid Zone Research Institute, Jodhpur (CAZRI) and Indian Institute of Horticultural Research, Bangalore (IIHR) or in the State Agricultural Universities (SAU's) particularly in Rajasthan, Haryana and Gujarat.

Realizing the potential of horticultural productivity or arid zones, the compatibility of these crops to the arid zone ecosystem and the need to achieve nutrition and income security for the people, the Indian Planning Commission, as recommended by the Working Group on Agricultural Research and Education, approved establishment of a National Research Centre for Arid Horticulture (NRCAH) during the Seventh Five Year Plan and upgraded to Central Institute for Arid Horticulture (CIAH) on September, 2001. Consequently, a Task Force, constituted by the ICAR in 1987, identified specific missions for its work. The Project Coordinator (Arid Fruits) was assigned additional responsibility of Officer on Special Duty to monitor the progress of the NRCAH in November, 1990. a site for its establishment was then selected and 125 ha land was acquired in November, 1992 close to the premises of Rajasthan Agricultural University, Bikaner (RAU) on national highway 15. Eight posts of scientists were redeployed from IIHR to provide scientific manpower to the NRCAH. In April, 1993 the Project Coordinator (Arid Fruits) alongwith the establishment of AICRP on AZF was lifted from Haryana Agricultural University, Hisar (HAU) to Bikaner to ensure the growth of the NRCAH which actually started on creation of the required administrative, technical and supporting manpower in September, 1994 "to conduct basic and applied research for improvement in productivity of horticultural crops and development of horticulture-based cropping system under arid environment; and to act as a repository of information related to arid horticulture", and to achieve this, the following objectives have been identified:

- The first objective is to introduce, collect, characterize, conserve and evaluate the bio-diversity of horticultural crops under arid environment.
- The second objective is to utilize the available bio-diversity and improve the target fruit crops such as ber, pomegranate, aonla, date palm and cucurbitaceous, leguminous and solanaceous vegetables to develop high quality and productive types having tolerance to biotic and abiotic stresses.
- The third objective is to study the factors related to rapid multiplication of propagules in case of established as well as new crops and the problems related to their growth and development
- The fifth objective is to study the eco-physiological parameters of cropping system models for utilization of high temperature and radiation resources.
- The sixth objective is to develop post-harvest technology package for extended use of the horticultural produce of arid region.

- The seventh objective is to develop integrated pest and disease management technologies for horticultural crops under arid environment.

Growth

Crop Improvement

Fruit Breeding Laboratory: This laboratory is to work on improvement of selected fruits or arid region such as ber, pomegranate and indigenous under exploited fruit yielding plants. The laboratory will also work on commercial crops such as grapes, citrus, guava etc. through collaboration with relevant Institutions.

Vegetable Breeding Laboratory: The laboratory is to work on improvement of vegetable crops of cucurbitaceae, Leguminosae and Solanaceae. Besides this, the laboratory will work on development of drought, heat and cold tolerant varieties for crops such as tomato, cauliflower, cabbage, pea etc.

Biotechnology Laboratory: This laboratory will work on tissue culture, micro-propagation and genetic engineering for improvement of horticultural crops.

Genetic Conservation Laboratory: This laboratory will work on conservation of genetic resources of horticultural crops *in vitro* and in field genebanks.

Crop Production

Fruit Culture Laboratory: This laboratory is to work on agro-techniques much as propagation, planting systems, pruning, fruit growth and development, orchard floor management, etc.

Vegetable Culture Laboratory: This laboratory is to work on nursery techniques, planting systems, crop rotations, interculture, off-season production, etc.

Plant Pathology and Nematology Laboratory

Water Management

Water Management Laboratory: This laboratory is to work on efficient use of water both under rainfed and irrigated systems.

Meteorology Laboratory: This laboratory will collaborate in research activities related to efficient use of water and climatic resources.

Soil and Plant Nutrition Laboratory: This laboratory is to work on integrated plant nutrient management.

Plant physiology Laboratory: This laboratory is to work on adaptations to stress caused by drought, cold, heat and salt.

Post-harvest Technology

Studies on Post-harvest techniques to improve storage life and quality of produce and on standardization of processing techniques of fruits and vegetables will be done.

Social Sciences and Extension

The unit will consist of Economics, Computer Centre, Statistics, Extension and AICRP Coordinating Unit.

Library

The library of NRCAH (CIAH) subscribes national and international journals and has a collection of a large number of reference books.

Research Farm

The research farm of CIAH has 124.58 ha land area. The farm has canal irrigation facility for more than 50 ha area. To ensure regular water supply, two water reservoirs with a storage capacity of 30 lakh litres of water have been provided.

Facility of two tubewells to ensure water supply to the laboratories, residences and the research fields in case of erratic water supply from the lift canal has been developed.

Buildings

The NRCAH (CIAH) has a provision for one double story Lab-cum-Office building covering a floor area of about 350m^2. A farm laboratory and farm sheds covering a floor area of 600m^2 have also been provided besides two security guard rooms, one at the gate and other in the heart of the farm complex. Seven residential quarters (1 Type V, 2 Type III, 2type II and 2 type I) have been provided.

Budget

The budget provisions on arid horticultural research started towards the end of V Plan which grew considerably from VII Plan onwards as a result of provision of an independent All India Coordinated Research Project on Arid Fruits followed by establishment of the National Research Centre for Arid Horticulture at Bikaner.

The details are given below:-

Plan	Period	NRCAH (Rs. In Lakh)	AICRP (AZF) Rs. In lakhs	Total Rs. In lakhs
Annual Plan	1978-80	--	11.02	11.02
VI Plan	1980-85	--	63.52	63.52
VII Plan	1985-90	--	124.64	124.64
Annual Plan	1990-92	15.00	89.89	104.89
VIII Plan	1992-97	350.00	225.00	575.00

Manpower

The manpower provision for arid horticultural research grew from 52 in 1978 to 158 in 1992.

Plan	Period	NRCAH	AICRP (AZF)	Total
Annual Plan	1978-80	--	52	52
VI Plan	1980-85	--	88	88
VII Plan	1985-90	--	103	103
Annual Plan	1990-92	08	103	111
VIII Plan	1992-97	55	103	158

Salient Research Achievements

The research achievements mostly emanating from AICR on AZF are listed below:

Genetic resources and crop improvement:

- Collection of Annona, Carissa, Cordia, emblica, Ficus, Morus, Phoenix and Ziziphus and cultivars/germplasm of Aegle marmelos (bael), Annona squamosa (Custard apple), Emblica officinalis (aonla), focus carica (fig), Grewia subinaequalis (phalsa), Pheonix dactylifera (date palm), Punica granatum (pomegranate) and ziziphus mauritiana (ber) are being maintained in field genebanks at different locations of India. Collection of Prosopis cineraria (khejri), Citrullus lanatus (watermelon), Cucumis callosus (kachari), Opuntia ficus-india (cactus pear) have been made.

- In ber, the early maturing cultivar, Gola performed well all over the country and the late maturing Umran had longer shelf life.
- In pomegranate, cultivar selection Jyothi and IIHR Selection from Bassein Seedless, G-137 from Ganesh and P-23 and P-26 from Muskat have been made. Some hybrids (*e.g.* Mridula) having better quality attributes have been developed.
- In aonla, besides the commercial cultivars Banarasi, Francis, Balwant and Chakaiya, new selections, *viz.*, Kanchan, Krishna, NA-6, NA-7 and NA-9 have been made.
- In bael, selection NB-5 has been found precocious and superior in fruit quality.
- In date palm, early maturing cultivar Halawy has been recommended. Some promising palms bearing large, non-astringent doka fruits have been identified from the seedling groves in Kachchh district of Gujarat.
- In custard apple, cultivar Balangar, Washington (PI 198797 and 107005) and interspecific hybrids Island Gem and Bullock's Heart were recommended. Hybrid Arka Sahan has been developed.

Production Techniques:

- propagation techniques have been standardized for ber (I-budding in polytubes), aonla (patch budding), bael (path budding) and custard apple (softwood grafting).
- High density plantation (5x2m) has been standardized in pomegranate.
- Pruning technique in ber (after 6 secndaries or 10-25 buds) and *phalsa* (at 100 to 150 cm height) have been standardized.
- Black polythene in ber, local organic materials like baddy husk and groundnut shells in pomegranate and *bui* (*Aerva tomentosa*) in date palm were identified for mulching

- Leaf sampling techniques have been standardized for ber (5th and 6th leaf from secondary shoots), pomegranate (8th to 10th leaf from tip), aonla (middle portion of 3 month old indeterminate shoots), custard apple (5th leaf from growing tip and phalsa (4th leaf from tip).
- Nutritional doses have been recommended for pomegranate (625g N+250g P+250g K) and fig (900g N+250g P+250g K).
- Fruit thinning in date palm (one-third strands from centre of bunch) and pomegranate (retaining 50 fruits per tree) and pre-harvest sprays in ber (750ppm Ethephon) and date palm (1000ppm Ethephon) have been recommended fro fruit quality improvement).

Crop Protection:

- Chemical control schedule have been developed against major disease such as powdery mildew in ber (2-4 sprays of 0.1% dinocap or carbendazim), black leaf spot of ber (3-4 sprays of 0.2% captafol or mancozeb or copper oxychloride or 0.1% carbendazim), fungal leaf and fruit spot in pomegranate (4 sprays of 0.3% mancozeb or 0.4% copper oxychloride or 0.1% carbendazim), bacterial leaf and fruit spot in pomegranate (3 sprays of 500 ppm paushamycin+0.2% copper oxychloride), Graphiola leaf spot in date palm (4 sprays of 0.4% copper oxychloride), rust in fig (4 sprays of 0.2% chlorothalonil or 0.4% copper oxychloride).
- Chemical control schedules have been developed against the major insect pests such as fruitfly in ber (prophylactic spray of 0.3% dimethoate or oxydemation at 70-80% fruit St.; 2-3 sprays of 0.05% malathion + 0.5% gur or sugar), fruit borer in ber (0.08% monocrotophos or 0.01% fenvalerate), pomegranate butterfly (one spray of 0.002% deltamethrin and one spray of 0.2% carbaryl 50 WEDP) and fig midge (2 sprays of 0.05% monocrotophos 40EC or fenitrothion 50 EC or 0.1% carbaryl 80 WP).
- Chilochorus nigritus has been identified as an important predator of *Parlatoria* scale in date palm.

Post-harvest Technology

Techniques for preparation of *chhuhra* and *pind khajoor* (soft date) from doka fruis (hard ripe yellow) have been developed.

Publications

363 research papers, 154 popular articles and 36 technical documents were brought out.

Techniques/technologies Generated

Thirty two agro-techniques were standardized at different centres in peninsular and north India. Three strategic technologies were also generated.

2

Impact Assessment

As a result of research for the last two decades, considerable knowledge in arid horticulture has accumulated and thus the sector has already assumed great importance for development of the vast drought prone arid zone of India.

There has been definite impact in terms of (i) awareness regarding potential of income generation by growing fruits and vegetables in arid region, (ii) increase in area and production of these fruits and vegetables in arid region and (iii) supply of fruits and vegetables for internal as well as export markets.

Growth of the Commodities

There has been spectacular increase in the production of fruits and vegetables in India during the past decade. The current annual fruit and vegetable production in India is respectively 39.5 and 70 million tones. In the arid region, the production of fruits increased from 0.6 million tones in 1984-85 to about 0.7 million tones in 1993-94 and that of vegetables from 0.6 million tones to 1.1 million tones during this period. Thus, even in the arid region the production increased rapidly. However, the availability of fruits and vegetables is very low in this region compared to the requirement of 1.1 and 1.9 million tones respectively considering the per capita per day dietar needs of 8.5g fruits and 250 g vegetables.

Fruits Vegetables

The growth of area under fruits and vegetables in the arid region has been rather slow considering the potential. Area under fruits increased from about 50 thousand ha in 1984-85 to about 66 thousand ha in 1993-94 and that of vegetables increased from about 63 thousand ha in 1984-85 to 85 thousand ha in 1993-94. Considerable area has come up under ber in different parts of the country. Its area increased from 243 ha in 1984-85 to 757 ha in 1994-95 in Rajasthan, from 626 ha in 1991-92 to 1432 ha in 1995-96 in Haryana and from 907 ha in 180-81 to 1108 ha in 1994-95 in Karnataka. Area under pomegranate also increased from 92 ha in 1984-85 to 370 ha in 1994-95 in Rajasthan and from 1418 ha in 1980-81 to 1746 ha in 1984-85 in Karnataka.

Area under vegetable increased from 48315 ha in 1984-85 to 66553 ha in Rajasthan in 1994-95, from 39249 ha in 1985-86 to 79500 ha in 1994-95 in Haryana and from 111628 ha in 1980-81 to 147683 ha in 1984-85 in Karnataka. Plantations of regular aonla, bael and date palm orchards have started coming up in different parts of the country. Area under date palm, however, has not increased to that extent owing to lack of a fast multiplication technique to true-to-type plants. The economics return potential from these fruits in now well known even from areas with precarious resource endowments.

Input/output Assessment

In case of ber under irrigated conditions, the benefit cost ration comes to about 6:1 whereas, in rainfed conditions, it is nearly 2.4:1. In case of pomegranate, aonla and date palm the benefit ratio is 2.4:1, 2:1, 3.2:1 respectively.

Fruit Crops

In ber, the identified varieties are in great demand. Tube method of propagation has been adopted by the nursery men for large scale multiplication. As a result, area under ber has considerably increased all over the dry areas of the country including peninsular region where it was almost non-existent earlier. The other production technologies related to pruning, moisture conservation by mulching, management of pests and diseases, etc. have not yet been fully adopted.

In pomegranate, the identified cultivars as well as the new hybrids are being planted by the farmers on large scale. The production technologies related to nutrient application, training and optimum fruit load, control schedules against insect pests and diseases, water use and moisture conservation by mulches have not yet become very popular.

In aonla, the identified cultivars are spreading in different locations in India. Propagation technique is already being used by the farmers. The cultural techniques related to planting on saline wastelands and nutrient application are being popularized. In date palm, the identified varieties have not yet spread in spite of great demand because the available multiplication technique is too slow. The cultural techniques related to fruit thinning and control of insect pests and Graphiola disease have not been adopted by the farmers.

In other fruits also wherever varieties have been identified, these are becoming popular. However, their regular cultivation in still negligible.

Other Horticultural Crops

Technologies specific to arid areas for other horticultural crops such as vegetable, medicinal, aromatic and ornamental crops are required by the farmers.

Lessons/constraints	Suggestions/options
Full exploitation of the potential of arid areas for the production of a large number of horticultural crops was not possible since techn?ologies specific to these areas were not available.	There is need to enlarge the programme of work to incorporate the fruit, vegetable, medicinal and ornamental crops which can provide nutrition security to the people along with economic and ecological sustainability to the region
The available biodiversity, which faces danger of erosion could not be properly conserved in the absence of repositories.	Well defined germplasm, repositories and genebanks should be established at appropriate locations.
Crop failures and productivity losses are common as a result of extremely high aridity leading to drought conditions.	A major programme should be directed to genetic improvement for drought and heat tolerance besides the study of physiological adaptation mechanisms to these conditions.
In arid environment, monoculture system is risk prone as well as less productive.	For sustainability and profitability, multistory compatible crop combinations need to be worked out.
The fruit crops which are already established in the region face some productivity constrains, e.g. i) Powdery mildew in ber ii) Leaf and fruit spot disease, fruitfly and fruit cracking in pomegranate	There is need to lay emphasis on integrated approach for management incorporating bio control agents.
Absence of standard varieties and/or multiplication technique for some indigenous/established fruit crops come in the way of their systematic production, e.g. Prosopis cineraria, Capparis decicua, date palm, etc.	Development of fast multiplication techniques is required utilizing micropropagation and tissue culture approaches.

Scenario

Vast Land Resource

The vast area under the arid lands is a great resource for the production of horticultural crops. The Indian arid zone is spread over nearly 31.7 million hectare land area of which 41.5 per cent is arable and 19 per cent is culturable wasteland.

Arid Regions in Different Zones of India

	Total area (m ha)	Arable area (m ha)	Culturable wasteland (m ha)
Rajasthan	19.6	9.20	4.20
Gujarat	6.2	0.02	1.66
Punjabdk	1.5	1.05	0.03
Haryanaa	1.3	1.15	0.02
Peninsular	3.1	1.63	9.11
Total	31.7	13.05	15.02
J&K	7.0	0.07	0.03

Indira Gandhi Canal Project Area: The region has a major canal irrigation programme to cover 15.37 lakh ha area under the Indira Gandhi canal Project having a total feeder canal length of 649 Km.

Genetic Resources

Indian subcontinent is an important gene centre being origin of nearly 166 cultivated plant species and rich in diversity of as many as 320 species and their wild relatives. The arid region has comparatively fewer plant species but several of them are of great horticultural value. Ziziphus mauritiana (ber), Z. nummularia (jharber), Cordia myxa (gonda) yield edible fruits. Seeds of Salvadora oleoides (pilu) yield oil. The seeds of Acacia senegal mixed with unripe pods of Prosopis cineraria (khejri), unripe fruits of Capparis decidua (ker) and ripe fruits of Cucumis callosus (kachri), commonly found in arid region, are used to prepare a delicious vegetable (panchkuta). Similarly, clusterbean, cowpea, amaranth, Cucurbitaceous crops are common vegetables. Fully ripe pods of khejri are very nutritious and contain nearly 56% carbohydrates and over 10% protein.

Euphorbia antisyphilitica and Simmondsia chinensis have performed well in the region. The region is also rich in medicinal plants such as Balanites aegyptica (hingota), Commifera wightii,

Catharanthus roseus (periwinkle), Withania somnifera (ashwangandh), Aloe barbadensis and Cassia angustifolia (senna).

Promising Germplasms

The native species such as Prosopis cineraria (khejri), Ziziphus nummularia (jharber), Z. mauritiana var. rotundifolia (boradi), Capparis decidua (ker), Citrullus lanatus (mateera), Cucumis callosus (kachari) have considerable variability. Fruit species such as Ziziphus mauritiana, Punica granatum, Emblica officinalis, Phoenix dectylifera (date palm), Psidium guajava (guava), Aegle marmelos (bael), Morus spp. (mulberry), Carissa carandas (karonda), Cordia myxa (gonda) have performed well.

Under the irrigated conditions, excellent quality production from fruits such as grape, sour lime, sweet oranges, kinnow, mango, papaya, vegetables such as cucurbits, beans, tomato, brinjal, chillies, okra, onion and leafy vegetables, medicinal plants such as Ocimum sanctum, O. basilicum, Solanum sp., Opium sp., isabgol, vetiver, liquorice, henbane and anise flower crops such as Damask rose, marigold, jasmine, etc.

Agroclimate Conducive for Quality Production: The peculiar agroclimatic conditions of the region imparts unique quality in fruits such as ber, sweet oranges, papaya and vegetables such as chilli, cumin, fenugreek and cucurbitaceous crops. Arid region is the only area where commercial date palm production is possible.

Surplus Family Labour: Owing to long slack seasons and low intensity of cropping, there is considerable surplus family labour available in the arid zone farming communuities which can be put to use for increasing production of high value horticultural crops.

Research Infrastructure

The National Research Centre for Arid Horticulture alongwith the All India Coordinated Research Project Centres located in 8 State Agricultural Universities and 2 ICAR Institutes now provides a strong infrastructure for research on various aspects of arid horticulture.

Weaknesses

Low Productivity of Horticultural Crops: Considering that the small area under horticultural crops in the region is mostly under assured irrigation and fairly good care, the productivity of these crops is low (Fruit 11.04 t/ha.; Vegetables 13.36 t/ha.). The production of the fruits and vegetables in arid region is respectively 0.7 and 1.1 million tonnes.

This is mainly because of inadequate production technology, related to adaptable genotypes and optimum management. Although the region has great potential for quality production of fruits such as grapes, pomegranate, ber etc., vegetables and other horticultural crops, the suitable package of practices for these crops to achieve high productivity under its peculiar agroclimatic conditions are not available.

Poor Soil Conditions

The arid zone soils are very poor in fertility. The soils of the north western arid region described as 'desert soils' and 'grey brown soils' of the Order Aridisols are light textured. The solum of these soils is moderately calcarious (0.2% $CaCO_3$) and below this solum at depths of 40 120cm, a sharply differentiated zone rich in alkaline earth carborates (5 45% $CaCO_3$) is present in the form of hard crystalline concretions which may be many metres thick. Most of arid areas (about 64.6%) is duny where the soils often contain 3.2 4 per cent clay and 1.4 1.8 per cent silt. The brown light loam soils occupy 1.7 per cent area which has loamy fine sand to fine sandy loam on the surface and a heavier subsoil underlain with calcium carbonate concretions. Besides this, about 5.9 per cent area is covered by soils having hard pan, 5.6 per cent is under hills and pediments, 6.8 per cent area is alluvial duny and 1.6 per cent is sierozems extending from the soils of Haryana and the Punjab. In Gujarat also, grey brown soils are widespread besides a large area having deltaic alluvium with small area in Kachchh having deep black and medium black soils. In the peninsular India, a considerable part of arid region has red sandy soil and some parts have mixed black soils.

The soils are poor in organic matter having per cent organic carbon of 0.03 in bare sand dunes to 0.1 in the stabilized dunes.

The soils are generally rich in total potassium (8250 18980ppm) and boron (1.9 12.2ppm) but are low in nitrogen (185 150ppm) and phosphorus (285ppm) and micronutrients such as copper (10 11ppm), zinc (2.07ppm) and iron (20ppm). The soils often have high salinity.

Poor Ground Water Resource

The ground water resource is not only limited owing to poor surface and sub surface drainage but is also generally highly saline. The depth of water ranges from 10m to as high as 140m. The other irrigation water resources in the region are seasonal rivers and rivulets in Gujarat, surface wells and some runoff water storage devices (*e.g.*, tanks, khadins). Thus the water resources in arid region are limited and can irrigate hardy 4% of the area.

Low Rainfall

The mean annual rainfall in the Indian arid region is very low and varies from 100mm in north western sector of Jaisalmer to 450mm in the eastern boundary of arid zone of Rajasthan. In Gujarat, it varies from 300 to 500mm and in Haryana and Punjab from 200 to 400mm. In peninsular region, the rainfall varies from 520 mm in Bellary (Karnataka) to 748mm in Cuddappa (Andhra Pradesh). Most of the precipitation in north western arid region occurs during July-September in about 19-21 rain spells. The moisture balance scenario in arid region is given below:

Moisture Balance Scenario in Arid Regions

		Rainfall (mm)	PE (mm)	Moisture index
Rajasthan	Solar radiation (cal/cm²/day) winter – 414 summer – 612	100-450	2063	-70 to –91
Gujarat	Temperature -4 to 50°C	300-500	2144-1700	-70 to –82
Punjab		200-400	1656-1362	-59 to -73
Peninsular India	Wind speed 12-16 km/hr	520-750	1859-1738	-60 to -71

Threats

The arid zone ecosystem is very fragile and is prone to serious imbalances even with slightest disturbance owing to mismanagement of resources or influx of external factors.

Unscientific Use of Irrigation Water: Indiscriminate use of water has already led to rise in ground water table and salinisation of soils in the region making large areas unfit for horticultural production.

Genetic Erosion: With the growing population pressure the indigenous genepools are getting eroded. Since this trend is continued to grow there is danger of extinction of native gene pools. Also most part of arid region is on international boarder, aggrevating such a threat for the region.

Recurrent Drought and High Potential Evapotranspiration: Arid environment poses constraints to productivity mainly because of abiotic stress owing to scanty rainfall, high summer temperature, high solar radiation and high wind velocity, low winter temperature, low soil fertility and high salinity in irrigation water and soil.

Potential Evapoptranspiration (PET)

The potential evapotranspiration (PET) in the arid districts of Rajasthan varies from 2063.2mm at Jaisalmer to 1503mm at Sikar with a moisture index value of 68.9 to 91. In Gujarat, the PET varies from 1700 at Surendranagar to 2144.6mm at Rajkot with moisture index value of 70 to 82%. In Haryana and Punjab, the PET value ranges from 1615mm at Hisar to 1362.8mm at Ferozpur with moisture index value from 58.9 to 72.9. In the peninsular arid region, the PET value varies from 1857.1mm at Anantapur to 1738.1mm at Bellary with moisture index value of 68.6 to 70.1. This region is characterised by the extremes of temperature having dry and severe winter with temperature as low as 4°C and very hot summer with temperature as high as 48 50°C. In summer, hot winds (loo) are common and sand storms are frequent. Wind speed can be as high as 12 to 15.8km/hr particularly during the summer. The solar radiation both in the northern and peninsular India is intense (from 414 cal/cm2/day in winter to 612 cal/cm2/day in summer at Jodhpur). Owing to these conditions, the atmospheric vapour pressure deficit reaches as high as 30mb during May June. These conditions get aggravated in the low rainfall years.

Animal-bird-insect Pests

The presence of animal insect bird faunal complex particularly concentrating in the localised green belts often poses problems. The disease occurrences are comparatively fewer but are often observed to adversely affect.

Exploitation of Indigenous Variability

Reserve of considerable variability of a number of hardy plant species which yield edible fruits and products of horticultural value provide opportunity for their exploitation by genetic improvement and systematic plantation. Besides, there is ample scope for introduction of useful plants from isoclimatic regions of the world.

New Introductions

Several fruit yielding species common in the arid region have not yet been commercially exploited mainly owing to the absence of their standard cultivars/types. Similarly, there is scope for production of plant species introduced from the isoclimatic regions of the world, *e.g.*, cactus pear, quandong, carob, African dove plum, argan tree, marula nut, nance, oyster nut, ye eb nut, etc.

Genetic Improvement

Genetic improvement in the traditional fruits such as ber, pomegranate, aonla and vegetables such as chilli, tomato and cucurbits to induce resistance against abiotic stresses with particular reference to drought and heat should further boost their production in the region. Some of the traditional horticultural plants suffer from productivity constraints owing to their susceptibility to specific disease and pest problems, *e.g.*, powdery mildew in ber and cucurbits.

Thus by genetic improvement in the traditional as well as presently wild but potential species and introduction of suitable exotic species, the major casual factors for the slow pace of spread of area under horticultural crops and for their extremely low productivity in the arid region can be mitigated.

Production for Export

Since arid zone agroclimate offers a great potential for production of high quality produce, there is great opportunity for production for export. The potential of quality production of citrus fruits such as sweet oranges and kinnow, ber, pomegranates and vegetables such as tomatoes, chillies and spice crops such as cumin, isbgol are well known.

Value Added Products

There is great scope for developing high value products from he indigenous produce such as Capparis, Cucumis and from the commercial horticulture crops.

Runoff Farming

Although soil conditions in arid region are poor but the characteristics such as coarse texture and crust formation on soil surface, respectively, cuts water losses through capillary movement and induces good runoff water yield even under low rainfall incidences.

Cropping Systems

The high solar radiation resource in the arid areas provides opportunity for its harvesting by optimation of cropping system models incorporating multilayer receptor crops.

Perspective

The current annual fruit and vegetable production in India is respectively 39.5 and 70 million tonnes which will have to be increased to meet the internal consumption need of about 50 and 143 million tonnes respectively by 2020 AD. The arid regions cover nearly 12 per cent land surface of India of which about 59 per cent (13 million ha arable lands and 6 million ha wastelands), given technology inputs, can greatly contribute in this endeavour. At present only about 66 thousand and 85 thousand ha area is under fruit and vegetables respectively giving a production of only 0.7 and 1.1 million tonnes as against current requirement of 1.1 and 1.9 million tonnes. By 2020 AD, therefore the production of fruits need to be increased three times and that of vegetables five times.

Export requirements of horticultural produce and products are expected to grow very fast in the post GATT scenario. The peculiar dry and warm blend in the arid region agroclimate offers opportunity for producing quality products of high health standard. This obviously would require R&D preparedness.

The population pressure would result in a) shrinking of agro resources and aggravated pace of erosion of the indigenous bio-deversity and b) increase in surplus labour particularly women. Conservation and characterization of the available genepools and valuable gene sources for their use for the benefit of mankind would require concerted efforts. To make the horticultural production system profitable and sustainable under the competitive post GATT scenario precise technology package needs to be developed incorporating efficient use of inputs for quality production under arid environment.

At present, commercial production is restricted to a small number of fruits and vegetables. Over the world only about 26 fruit yielding species are commercially grown whereas there is a reserve of over 3000 of which at least 200 have economic potential. Fast growing information technology and modern communication networks have already increased awareness in the world communities about the rich variability in exploitable horticultural plant species. Some less known fruits and vegetables have already attracted the palate of the people in distant corners of the globe. Fortunately the Indian subcontinent is a rich reservoir of such resources. Arid regions also abound in 24 new plant types.

Recurrent droughts and persistence of extreme aridity conditions result in crop failures so common in the region. A considerable land area is degraded. If left to itself, the growing biotic pressure and abiotic stresses would further aggravate the already precarious conditions. Appropriate and satisfactory safeguards to these conditions in the production system will have to be developed to make it sustainable and profitable. Productivity enhancement through genetic manipulations and by efficient management of natural resources and inputs such as water, nutrients and soil will need greater attention. Amelioration and profitable use of the degraded lands would also assume importance. Monoculture production system is often unprofitable owing to less efficient resources utilization and risk of climatic calamities in the arid region. Therefore, developing appropriate crop combinations doughtailing with the available agroclimatic resources would assume great significance.

Owing to their perishable nature value addition by suitable processing techniques will be essential for stablilising income from horticultural crops. Also exploitation of income potential from the indigenous produce can be enhanced by value addition. Technologies needs to be developed for horticulture enterprise such as nursery, post harvest handling of produce and processing for gainful employment to the rural people.

The relationship between the existing animal bird insect fanual complex with the horticultural plants often create problems in their intensive production particularly under the limited resources of the arid region. Therefore, an integrated management systems will need to be evolved for sustainable production. Similarly, integrated management system for the relevant diseases will also need to be worked out.

Therefore, the approach has to be to:

(i) explore and conserve the fast eroding indigenous plant genetic resources of fruits, vegetables and other horticultural species and to introduce useful plants from isoclimatic regions of the world.

(ii) improve productivity of horticultural species such as ber, pomegranate, aonla, Prosopis cineraria, Citrullus sp. and Cucumis sp. for growing under conditions of the prevailing abiotic stresses,

(iii) Optimize use of the scarce water and high solar radiation resources production of horticultural crops.

Wasteland Horticulture

In India, the wastelands occupy about 53 per cent (175 million ha) of total geographical area. Most of this area is under arid and semi arid desert. The peculiar conditions of these areas such as abundant sunshine not only provide prolonged growing season but also make it possible to have multi storey cropping without disadvantage to interplants and ground crops in their photosynthetic efficiency. The rolling topography of these wastelands also offer opportunity for runoff generation and efficient use of this scarce input for maximum productivity. The other wastelands such as undulating and rocky areas, pediments, piedmonts, saline lands etc. also call for systematic development for generating horticultural production in arid areas.

Identification and development of straterigies for production/ exploiting existing genetic variability for commercial production of fruit/vegetable crops under these conditions can provide profitability for sustenance to the people inhabiting these lands.

Production of Planting Material

Technologies will be developed for large scale multiplication of seed/planting material of standard varieties/cultivars of plant species suitable for arid region.

(a) National Institute such as Central Research Institute for Dryland Agriculture, Hyderabad, Central Soil and Water Conservation Research and Training Institute, Dehradun, Central Soil Salinity Research Institute, Karnal, Water Technology Centre of Indian Agriculture Research Instt, New Delhi, Directorate of Water Management Research, Rahuri.

(b) International Institutes such as Internaltional Crop Research Institute for Semi Arid Tropics, Hyderabad, University of California Davis, California, USA, Faculty of Agriculture, Rehovot, The Hebrew University of Jerusalem, Israel, Utah State University, Logan, Utah, USA, Irrigation Research Division, Commonwealth Scientific and Industrial Research Organisation, New South Wales 680, Australia.

Execution and Coordination

The execution and coordination of the programme will be effected by the Director, National Research Centre for Arid Horticulture as per its mandate and priorities fixed from time to time by the Division of Horticulture, ICAR and the relevant Committees of the ICAR through the instruments of the

Management Committee and Staff Research Council. The programme will be implemented by the different research laboratories supported by technical and other units as illustrated below:

AICRP on Arid Zone Fruits

Research on fruits like ber, date palm, pomegranate, fig, custard apple, etc. was initiated on adhoc basis since 1976 under a AP Cess Fund Scheme entitled "Research on some selected fruits in the semiarid areas in India" at ten centres. During the sixth Five year Plan, this adhoc scheme was merged as such and was named as the Cell III of the AICFIP. At present, there are 12 centres functioning under the Project.

The technologies generated by the NRCAH will be tested on all India basis at the centres of the AICRP on Arid Zone Fruits so as to arrive at usefulness and applicability of results at different locations in the country. The AICRP on Arid Zone fruits will also continue to conduct multilocation testing of the technologies generated by different institution/organisations in the country which are conducting research on the fruits relevant to these areas.

Risk Analysis

The research strategy of NRCAH will be directed towards solving problems in horticultural productivity in backward areas of the arid region which is well known to have uncertainties in productivity of largely rainfed crops. Even in irrigated micro pockets, the potential productivity diminishes under the extreme conditions which are of recurrent occurrence. The drought epidemics are common. Therefore, germplasm conservation will have to take care of these risks besides the biotic interferences by wild life and owing to proximity to international border. The problems related to drought, erratic rains and extreme radiation will have to be considered in experimental procedures to be adopted.

Output and Expected Situations

It is hoped at the end of 2020 centre would generate technologies for horticultural crops of arid region so that internal consumption of fruits & vegetables for the in population would be met from local production as well as would be able to earn revenue from exports.

Project Review, Reporting and Evaluation

Present NRCAH is part of the Division of Horticulture headed by Deputy Director General (Horticulture) in Indian Council of Agricultural Research under the Department of Agricultural Research and Education, Government of India, new Delhi. A Management Committee will oversee and catalyse the Centre's activities and a Research Advisory Committee and a Staff Research Council will advise on its research needs, priorities and approaches through periodic meetings. The research programmes and activities will be constantly monitored by the Division of Horticulture of the ICAR.

The results of the work done will be published every year in an annual Report besides through other publications. The research needs, programmes and approaches will be reviewed every five years by a Quinquennial Review Team constituted by the ICAR.

Resource Generation

Consultancy could be provided for field application of the technologies through organising training programmes to development functionaries of Government and non governmental organisations as well as interested growers. Beside this consultancy for organising seed and plant multiplication units establishments of intensive production units for field productions as well as post harvest utilisation could be provided in due course.

Orissa Vision-2020

Towards Building a New and Modern Orissa

Synopsis: Orissa is more known for many wrong reasons. Whether it is starvation death or selling of small kids or ugly political scene by our inefficient politicians, Orissa has become a needy state on the part of international organizations a laboratory state to experiment with poverty. Where are our

Oriya intellectuals? Oriya intelligentsia's approach of grabbing opportunity and habit of shifting residence to other parts of India by totally forgetting their native state and people is a negative development so far Orissa is passing through.

Orissa is also not expecting from them contributions financially all the time, rather what is required is will power, ideas and determination. I think, everybody will agree with me that we have lost everything in terms of self-respect. To make a difference, all of us have to take a new start by promoting Orissa in a big way and we have to review our own work and our contribution to Orissa. I do not think that we are short of talent but what we need is entrepreneurs to make use of those talents.

Have you seen any person, who has become successful without a vision and plan? Orissa is suffering from this chronic disease and moving without any long-term plan and vision so far. Let's start dreaming and you will find yourself in a vision, which will take you to a focal point, and unless we focus on action and implement it too, there will be no result.

So, vision has a power to change anybody's fate. The people rise up, only when crisis compels. That is what is happening to our Oriya intellectuals? When there is a flood or drought, just collect some relief and send with some suggestions. Is this any way helping to solve this problem? As we have seen, Orissa is displaying all typical symptoms of a failed state and the culture of non-performance being the hallmark; we must start the initiatives outside the realm of government involvement. I am not advocating to ignore the government but should not wait for government to solve all our problems.

When we look the data sets for Orissa, any body will be shocked. Orissa has the highest infant mortality in the country. Our two-third of the rural population is living in abject poverty and health facility is pretty bad, as we have lowest number of doctors per capita in the world. Most surprising thing is that we have less than 5% of the population having access to subsidied food aimed at poverty alleviation. State's economy has incurred huge losses to the extent of Rs 13,230 crore due to natural calamities from 1998-2003. More than 90% of the state plan outlay is being financed from borrowings. Orissa has the highest proportion of population living below poverty line. Around 47.13% (all India average is 26.1%) in Orissa lives below poverty line. You will be astonished seeing the average of BPL in Jammu and Kashmir, which is less than 3.5%. And there are wide disparities between coastal and southern regions. It is because of high concentration of Scheduled Tribe population in the southern region. Scheduled Tribe population in coastal area is 7.2% as against 39.7% in the southern region. So, poverty ratio in coastal region is 64.0% as against 85.5% in the southern region. If we see the overall trend, we find that 80% of the rural families are leading a painful life in the state. And we lost many investment opportunities due to non-availability of basic infrastructure in our capital city of Orissa. Bhubaneswar is a dead city for outsiders and it is the real worry for investors to start something new in Orissa. Observing all these mess, I thought of collecting the ideas of our intellectuals and make them involve in sharing the suffering process of Orissa.

Where is the slogan of "One thousand industries in one thousand days" of government of Orissa gone? Starvation death and child sale is making sensational headlines. The poor of Orissa ravaged by cyclone, flood, drought and famine has become the bread and butter of numerous voluntary agencies. Our local media has adopted a style of negative reporting, largely controlled by politicians with their own agenda. The culture of cynicism has grown rapidly infecting the psyche of the entire population of Orissa. The young people are struggling for maintaining themselves, as there is no job for anybody in the state. Our past leaders thought of to make Orissa prosperous through mega projects sponsored by central government. But general mass have benefited very little. If instead, they have thought of establishing technological institutions to educate talented Orissan, then, we have not in this situation. Building of an educational atmosphere is most important for any kind of development. This is one such attempt by the contributors of this volume to highlights the problem from their long time involvement in the particular field.

The whole document is divided into four major sections. The first section is having more than twenty-thoughtful articles, second part deals with three discussion papers on the topic of power sector reform, education system in Orissa and Pachima Orissa Vikash Parishad and the idea of introducing a section like this is to respect divergent views expressed by several scholars on a specific topic. Then the third section is having a few short comments by a few eminent scholars on the aspect of underdevelopment and poverty. The fourth section is having a database on Orissa with all-important human development indicators. Apart from these regular sections, the book also has a thoughtful preface and current introduction section and finally, a section called another Orissa is possible, where the editor tries to argue certain trends, which Orissa requires immediate steps to take Orissa to a comfortable position. Here, I will try to bring the synopsis of all the articles in an integrated manner rather than conventional way of describing all articles separately.

The authors have expressed their concern in the following ways with some concrete suggestions-like one paper talks about uncontrolled increase in population and poor management of the ecosystem has lead us to an Orissa where urgent actions are needed in all fronts to restore the lost natural resources. It is disheartening to notice that about 80% of the saplings/plants planted die because of lack of water. However, a relatively new technology called "Gel hydroponics" will make the plantation endeavour amazingly simple and successful. Gel hydroponics, by making plants drought tolerant, increases the survival potential of the saplings/plants. Effectiveness of drought tolerance through gel hydroponics is a direct function of gel quality, which is reflected by its water holding capacity and durability in the soil.

One of the papers highlights how badly natural calamities affected the State's economy in 1999 due to heavy cyclone in the coastal belt. Similarly, the insufficiency of rainfall in the 2000-01-crop season resulted in crop failure and drought situation in large parts of

Western Orissa: The recent flood during 2002 affected the state's economy severely. The overall poverty ratio, taking both rural and urban areas together, was also the highest for Orissa among 25 States during 1987-88. During 1993-94, the position improved marginally and Orissa occupied a position above Bihar in respect of poverty ratio for rural areas as well as for the overall poverty ratio. So in this sense, water projects can bring about affirmative changes in the lives of the poor people by overcoming poverty, improving health and well being, and by enabling individuals and groups to act to develop their lives.

One of the author has analysed the recent development in agriculture and explained the advantages arising from, the National Agricultural Technology Project (henceforth NATP), sponsored by World Bank, which has come up with a mega technological package for the development and dissemination of agro-based technologies.

Operationalizing its determined objectives, NATP follows a multi-agential approach by establishing a network between the public and private agencies along with the involvement of 'Third Sector and also the farmers'/women farmers' organizations. The current policy framework in the field of agricultural research and extension through the NATP makes an effort to accomplishing the decentralizing decision-making process; public-private interface; skill up-gradation; farmers'/ women farmers' empowerment; mainstreaming of gender; and emphasises on information technology.

Orissa is blessed with five important resources says another author like, Natural Resources, Mineral Resources, Agricultural Resources, Biological Resources and Human Resources. However, Orissa has gaps in entrepreneurial developmental congruence between people, ideas, government and financial systems. According to the paper, Orissa needs a visionary development campaign, developmental networks, linkages between public-private systems, and more importantly effective management of its resources and then it can rise to the challenges and opportunities in many sectors.

Few papers campaign for setting up of industries in sectors like aluminium, sponge iron, granite cutting and polishing, telecom cable, steel, petrochemicals, fertilizer power, sugar, chemical and pharmaceuticals, telecom etc. For this, private initiate and private investment need to be encouraged.

Mere construction of *panthasalas yatriniwases, panthikas, aranya niwas, panthaniwases* does not in itself carry the potential to attract tourists says one of the papers. For tourism to excel as a revenue-generating industry, Orissa must re-design, renovate, and beautify our places of tourist and religious importance to a great extent says one of the authors.

Adding an artificial, aesthetic as well as a post-modern appeal to our existing sports having tourism potential is needed argues one of the papers. There is a need to rescue the temple from the domination and *ghettoization* of the *Pandas. Mono-departmental approach should* be replaced by a *multi-departmental approach*.

That is, line-departments such as Horticulture, Animal Husbandry, Fishery and Soil Conservation should be made to work collectively for a holistic intervention, which is lacking in the state. The threats like siltation, loss of salinity, extensive weed growth and depletion of fishery resources ought to be tackled sternly. Prawn culture in the lake be regulated without jeopardizing the concerns of the traditional fishing communities. There is a need to rescue Chilika and differentiate between the monopoly of the alien mafia dons and the livelihood needs of the locals. To get out of the mess from the present educational system, one paper argues that the "autonomous" status accorded randomly to colleges should be withdrawn urgently so that the monopoly of teachers and the subsequent corrupt practices can be checked and most vitally, the quality of teaching can be maintained. Every college should not be given the power to function as mini-university in every town. A mobile squad team can be created to pay surprise visit in order to inspect classes/ teaching and the college functioning. Posting to the capital city or other well-equipped mega cities should not be a monopoly of select academicians (referred to as *chamchas* of politicians/ bureaucrats), rather no person should be allowed to continue for more than five years in a particular city.

The State Government should enact a *Gender Equality Act* to offer a special concession to women in work sphere. Under this, women should be entitled to take a day's holiday every month during their menstrual period; those who suffer miscarriages can take seven days to four weeks off. Few of the articles reflect upon the coastal-western divide and its long-term impact upon the development of the latter. One author terms it unfortunate the coastal-biasness of the administrators in the setting up of developmental milestones. The recent decision to found an AIIMS pattern hospital in Bhubaneswar is a case in point. Instead of choosing a poorly health-cared area, the government has unfortunately selected Bhubaneswar, which is already having abundant as well as high-tech health care facilities.

Poverty, regular natural calamities and lack of infrastructure have been a general issue of concern for *Orissa's backwardness* in modern industrialization mentioned in one of the articles. In contradiction, *Orissa* has advantages of mineral rich reserves, large coastlines, great tourism potential, and low cost skilled and unskilled work force. Therefore a clear vision of what can be achieved and how best to achieve it, is required for the state. But, *Orissa* has a shortage of high-qualified human resource. *Orissa* is still one of the lower graduate producing states compared to other Indian industrialized states like *Maharastra, Karnataka, Tamil Nadu, West Bengal* and *Andhra Pradesh*. A modern approach for infrastructure development in basic science and technology is still required for the state.

Foundation for a strong higher and technical education as well as research must be built inside the state, during next five years. Few more new universities, research centres and technical colleges should be established in the state. Importance should be given to the backward regions for higher infrastructure development.

There is a consensus among the writers regarding the importance of English education in Orissa. As English is very important for globalization, good education in English and Oriya languages along with special emphasis in spoken English should be given importance at the basic level.

Construction of four line highways is necessary for most of the busy roads in the state, *for example: one along the coastlines, one in the western border line, and one in the heart line of Orissa have been suggested in the paper*. Simultaneously, government should give special attention to build new railway infrastructure in the state.

Echoes of the philosophy of self-reliance can be found in the growth of the open-source software (OSS) movement, which probably had its genesis with the GNU project in 1948 in the United States of America. Many, including President Abdul Kalam, have described the technical and social benefits of OSS say one author. He told that the original metaphor that we started with, besides the *khadi* that the *charkha* produces, the Mahatma also promoted it as a symbol of self-reliance; and the daily act of spinning yarn as a tool to focus one's mind. Therefore, in that spirit, It offers you a discipline for the electronic *charkha* to spend at least half-an-hour each day.

One paper argues that CPR area as percentage of total geographical area of the state has significantly decreased from 20.39 per cent during 1970-71 to 15.54 per cent during the period 2000-01. The major factors responsible for the decline of CPR area are: legal privatization undertaken by the government under the scheme 'land to poor'; illegal encroachments due to higher growth of population and poverty. So vision document argues that the NGOs should develop a new and different partnership with the policy makers and government agencies in concert with progressive thinking on natural resource management. It is observed that the irrigation projects need to be revitalized and maintained properly so that water requirement of the agricultural of the state can easily be met.

Besides, the Government of Orissa should revive the tank irrigation system in the KBK districts to resolve the problem of water scarcity for the agricultural use as these districts which very often face scarcity of water even in the rainy seasons and experience regular droughts. There are four critical issues, which need to be looked at for the success of the community based watershed management project. These are community participation, sustainability, and management of community conflict and, most importantly, creation of the regional and local linkages.

One paper, highlighting the situation of unemployment in the state offers two models to solve the problem in the context of Orissa. One author says that we can club young people with similar minds and link-up for a particular type of work, add up their individual funds and the collective amount should find some way out for solving the problem. Small-scale set-ups like mixture factory, selling different products in distant villages around a small town, setting up tuition and coaching centres etc. are viable options. There is no need for huge funds for these activities. Rather, if people come-together these will be boosting employment opportunities. Presently collective approach is the best option to solve this universal problem. Youth should make things in-groups and think about themselves collectively. Youth can start with small investments in the form of cooperatives and then build a resource centre on their own. People prefer government jobs because of the stability factor. In case, we are able to maintain stability for the product youth are supposed to produce (or nearly stable) then, one can certainly depend on the same, even collecting the garbage can provide a stable income to many youth.

He also proposes to establish a national level vocational training campaign which can be started with people who are well established in different business and professions nation-wide, become members and readily train the youth registered with us in their organizations. The salaried members can also cooperate by taking some time out in giving services in training the youth. For *e.g.*, a person employed in a software firm can very well take classes on software for people who wish to learn computer and which will help them get a job or to start a small firm.

One of the papers taking Orissa as a case explains how e-learning can help the state. Learning promises to provide a single experience that accommodates the three distinct learning styles of auditory learners, visual learners, and kinesthetic learners. Other unique opportunities created by the advent and development of e-learning are more efficient training of a globally dispersed audience; and reduction in publishing and distribution costs as Web-based training becomes a standard. E-learning also offers individualized instruction, which print media cannot provide, and instructor-led courses allow clumsily and at great cost. In conjunction with assessing needs, e-learning can target specific needs. And by using learning style tests, e-learning can locate and target individual learning preferences. The author argues for e-learning system for Orissa.

It is found that the poverty in Orissa is more concentrated in the tribal belt. Few of the write-ups stress on the development of tribal people in a sustainable ways. One paper argues that the failure on the part of scheduled tribe to cope with socio-economic growth and technological advancement can be attributed largely towards poor literacy rate among this marginalised group of people, which directly affects their health, hygiene, overall family structure which hinders social upliftment. Ignorance, lack of interest and lack of knowledge regarding the benefits of being educated and how far in life it carries a person is one of the reasons for parents not allowing children to go to school or inculcating the values of education in a child. Lack of Tribal Oriya bi-lingual teaching aids and books to bridge the gap of knowledge is another cause. Lack of teacher training institutes is another limiting factor. A part of the tribal child education task force should be involved in developing user-friendly *i.e.* tribal child friendly learning material. This should be followed by a group of the taskforce being trained to teach and another group trained to implement the project. They should also be suitably remunerated.

The point is that we have to build a network of roads so that these places can be put on the tourist map of India. The construction of roads will not only help to improve the sick tourism industry but it will also help to provide employment to the local people.

Touching the transport problem, one paper points out that the number of trains connecting main cities of Orissa are too less. For example, there is only one daily train from Mumbai to Bhubaneswar although the train is always packed to its capacity even in off seasons. This particular train takes about 39 hours to reach Bhubaneswar as it goes through most parts of AP. So, if a person wants to go for vacation why he will come to Orissa. One paper also touches the KBK region in depth. In Kalahandi, with 87 per cent of the population somehow surviving below the poverty line, the struggle to eke out a living is an onerous task, according to the author. More so, considering that the abundant resources, including land and forests, are owned by the "gauntiya", the feudal landlords.

For the poor, human bondage and surviving on the charities of the moneylenders, is perhaps an inescapable route. The innocent, simple and illiterate tribals end up borrowing money at the time of distress only to find themselves slipping deeper and deeper into debt. Although the moneylenders vehemently deny, the annual rate of interest comes to about 460 per cent. He argues to take note of the above things while preparing any policy for this area. If development assistance is the lone measure of economic growth, Kalahandi has failed miserably. Not even 10 paise out of a rupee of development aid has percolated down to the people who needed it most. The real beneficiaries are the government officials, who more often than not bribe their way to stay put somewhere in the four districts of western Orissa. At the same time, Kalahandi region has perhaps the highest concentration of NGOs and yet the poor continue to go to bed hungry.

One author feels that each of us can play an effective role if we can adopt the philosophy of "Pay It Forward", the film in which the young child favours three persons on the condition that they pass it forward to three persons each. In our respective fields, be it in Govt. service, business line, scholarship/fellowship, or in some other area, if we can help three Oriyas every year with a similar condition, it will work wonders. The paper argues that there are instances, how the NRIs of Kerala help other Malayalees to go to Gulf countries and the Punjabis help other Punjabis to go to UK or Canada. The Tamilians help other Tamilians to go to Malaysia and Far East.

Studies consistently found that the poor households pay a significant part of their income for health care, even when the government theoretically providing free or nearly free services, says one paper. Many patients have to bankrupt their family to pay for the expensive services or forgo the treatment. One of the papers explains why there are numerous canals getting clogged each year. This Victorian method of unclogging canals each year should be abandoned, the author argues.

In fact the whole network of irrigation budget needs a strict and urgent review. The same paper also highlights that another bigger task at the grassroots level is the education of parents (!), in child caring. Why is it that, on asking, most of my friends remain dumb over Odish A History? First of all, we must realise the fact that giving grants, building a school here and a hospital there can solve the problem at micro level for a micro amount of time-scale. But grants and aids will only make us perennial beggars. And more importantly, we can make a micro change in a macro scale (in time and effect and spread) or a macro change in micro scale. And, the real vision should be for the former one.

In a state like Orissa, where poverty level is very high and a large section of the people in the rural areas still depend a lot on forest resources, bamboo can prove to be a successful window to poverty alleviation, explains one paper. For this, one needs to go beyond the traditional planning domain and look into new and newer uses of bamboo.

Apart from that the concept of NGO village, talent hub, one hour internet connection, a regulation for MLA to be in the constituency rather than in the city are some of the suggestions found in the section another Orissa is possible segment. Hope, this document will act as a torchbearer in Orissa's progress by 2020.

Development of Backward

In the Himalayas the Committee accepts the areas already demarcated as hill areas of U.P., West Bengal and Assam as backward hill areas for special consideration. The exclusively hill State of Jammu & Kashmir, Himachal Pradesh, Arunachal Pradesh, Manipur, Mizoram, Nagaland, Meghalaya, Sikkim, and Tripura are already treated specially as back ward States and the needs of the backward areas in these exclusively hill States should be looked after in the State Plan on the basis of the guidelines given by the Committee for dealing backward areas. Excluding the areas covered under tribal sub-plan for which separate provision exists, the rest of the hill areas above 600 meters contour in the Deccan belt should be considered backward hill areas for the purpose of special development.

The programme for hill areas should be an integrated development programme for building up of infrastructure and for supplying a package of inputs and other necessary services almost simultaneously. Each component of the programme has to be inter related and coordinated with the other components, with the ultimate aim of maximising production and benefiting the hill people economically.

Strategy for Development of Backward Hill Areas

The strategy for development should be dictated by considerations of:

(i) Benefiting the people as a whole rather than the status groups;

(ii) developing local resources and utilising local talents so that the need for out-migration of the adult males is no more necessary, or at least gets considerably reduced;

(iii) transforming a consumption-oriented economy into an investment-oriented one;

(iv) regulating inter-regional terms of trade in a manner that they cease to be exploitative for the backward regions;

(v) the States filling up the gaps in the marketing mechanism to give the backward areas fair return for labourers; and

(vi) maintaining the ecological balance.

In order to slow the progress of outmigration, generation of gainful employment locally has to receive a high priority.

Methods of reducing drudgery in the women's life have to be found so that the women can also become participants in the development process and yet remunerated for their labour in the time saved from drudgery.

Gearing up of the production system and creating proper distributive machinery are from this point of view, the two primary requisites of ensuring flow of benefits of growth to the people.

In backward hill areas, self-sufficiency cannot be and should not be the goal of the development policy; but maximizing the income from each unit of the development base, either land or water or domestic animal. Therefore, a policy of promoting spatial specialisation needs to be favoured.

The problem of food can be solved if in areas wherever the food crops are replaced by new cash crops if:

(a) suitable marketing facilities are built into the system to buy the production at fair prices and take them off the hands of the farmers of the areas, and

(b) arrangements are made to supply the food grains commonly consumed in the area in fair price shops so that the farmer can buy his requirements at reasonable prices. An area approach of development will have to be taken up first in the zones which can be much more productive under the 'new strategies and on the communication routes'.

It is suggested that:

(i) necessary know how and facilities be provided to ensure more efficient utilisation of available resources, particularly those of the forests, cattle, land and water;

(ii) the framework of focal points be used for a more realistic and a rational location of various services connected with education, health etc.

(iii) resources should not be frittered away by unconnected developmental schemes but used for an integrated rural development approach on an area basis, with emphasis on production and better investment by the State on the necessary and infrastructural base for the production strategy.

It is recommended that the watershed should be the primary planning unit, rather than a block or a district. The watershed approach provides a sound basis for programming of soil conservation, water harvesting and harnessing and land use and planning social institutions. It is also possible to delineate watersheds into micro and mini watersheds. A mini water-shed may be equated with our concept of a focal point and may cover an area of a cluster of villages with population coverage of a 5000 or 6000 of a block population and micro watershed may cover an area of 1000 to 1500 hectares.

The strategy of entrepreneurial and skilled lab our development that the has recommended in its report on 'Industrial Dispersal' is all the more necessary in the backward hill areas. The Committee would, therefore, recommend necessary action to super-impose the industrial development in these areas by following guidelines in that report.

Industrial development in the hill areas has to be based essentially on the promotion of such activities in which hills offer a distinct advantage like forest based industries, vegetable and fruit processing etc. Another area in which hill areas offer an advantage in terms of climate and dust and pollution free environment is electronic and precision goods. The major advantage of such industries is that they do not involve heavy transportation of raw materials. These are mainly skill based industries where the value added is substantial. Necessary steps will have to be taken for training and skill development so as to prepare the environment for location of such industries.

Livestock has attained great importance in developing hill economy. Therefore, priority should be given for improving quality of livestock on commercial lines through cross breeding with exotic breeds. Pasture development programmes should be taken up along with measures to improve cattle.

Social forestry on scientific lines is necessary to check soil erosion, maintain ecological balance and produce industrial new raw materials, conservation, extension and productive exploitation of forests has to go hand in hand to meet the industrial requirements and general needs of the people.

The administration is generally not committed to programmes of backward hill area development because most of the time, people do not choose to serve in these areas. Posting in the hills is generally seen as a form of punishment. Therefore, officers charged with development programmes should have feeling of commitment to these areas and be sensitive to the needs and aspirations of the people.

The sub-plan idea is conceptually a sound one and must be regarded as a breakthrough in our area development concept. The underlying idea is one of achieving internal consistency and integration among various on-going programmes. It is an appropriate method of imparting some new orientation in area planning by taking note of the missing elements in the normal sectoral programmes of the State Governments.

While planning for big projects in under-developed hill areas, it would be necessary to build in adequate safeguards to protect the environment on the one hand and the interests of the local population on the other by enabling them to avail of the job opportunities that arise.

It would be necessary to work out suitable mechanism to replace the non-descript and unproductive livestock heads by introducing remunerative and fewer heads. It would be desirable to work out appropriate land use patterns so as to ensure adequate feed and fodder for improved livestock population.

Attempts should be to aim at appropriate land-use management in the hill areas, suiting to the specific site conditions, which would include wet cultivation and dry cultivation, horticultural and cash crop plantations and cultivation of utility trees/shrubs.

Watershed approach would mean planning and implementing soil conservation programmes for all types of lands and associated drainage system of selected watersheds within a reasonable time frame with the objective of providing maximum protection to existing land and water resources while optimising their use for increasing production and employment benefits.

In order to comprehend the total problem vis-a-vis the total potential and to draw up a programme which could be implemented within a reasonable period (say five years), the size of such watershed should be between 1,000 to 5,000 hectares. Once delineation and modification works are over, inters priorities and development of these watersheds are to be assigned taking into consideration the corrective factors for erosion and sedimentation ranging from physiographic to climate and including proximity to the reservoirs. The Central and State Soil Conservation and Soil & Land Use Survey Organisations would have to take up this task to identify, at the earliest, watersheds on a priority basis for preparation of integrated management plans and implementation of the same.

Watershed management planning will need data pertaining to watershed characteristics such as topography, soils, geology, vegetation and climatic data, besides incidence of erosion, flood and drought hazards. Therefore, there should be an adequate mechanism at the level for collection of such data through field State Headquarters and at field/units or from concerned agencies. The need for the basic data should be fully recognised and their collection, analysis and interpretation should be considered as a pre— investment towards proper planning and implementation of watershed management programmes.

The most logical step would appear to be to incorporate corrective measures in the existing land use system to make the existing land-use practices less vulnerable to erosion and degradation, hazards.

Alternative management practices should be introduced slowly to encourage the beneficiaries to shift gradually to the improved land-use pattern. If the agro-silvicultural practices or agro-horticultural practices can be made profitable while practising mixed farming, it may be possible to achieve this gradual shift in land-use pattern and retiring the steeper slopes from cultivation of common agricultural crops, to productive and remunerative fodder or tree cover.

Once priority watersheds are identified and land classification completed it will be necessary to identify the areas needing treatment under agricultural, forest and other land use practices. Depending upon a number of factors, such as slope, soils, rainfall, etc. Package of treatment measures will have to be chosen. These measures would include bench terracing, outlets with drop pits, riser protection; rehabilitate deteriorating terraces, hill side ditches on the agricultural lands.

For pasture lands combination of grasses, legumes and fodder trees and bushes will have to be chosen. Closures and erosion control measures like dykes may have to be adopted here as well.

Water harvesting structures such as low canal sills, diversion weirs and some pons in suitable locations, could be considered for appropriate utilisation of available run off. Check dams, revetments and spurs will be necessary for controlling gullies and erosion along the stream banks and beds as well as land slips along the roads. Some arrangements will need to be made in the planning Cell for collecting representative base level data for typical areas representing dominant combination of practices such as bench terracing, afforestation, water harvesting system, pasture etc.

There are practical difficulties in mobilisation of self-labour in innumerable individual holdings in a single terrace for subsidised working through extension methods under the Soil Conservation Acts. Land records need updating. All this has resulted in the concept of integrated watershed management being rarely achieved, particularly when a large number of farmers are reluctant to join in the working. At the same time, any soil conservation programme would be self-defeating if the people on whose lands these are carried out are not only involved in it effectively but have some stake in improving the land and maintaining it. Considering all this, the Committee would recommend:

(i) that the existing practice of subsidising private works on farmers' lands to the extent of 50% should be continued.

(ii) if there are any works on the private lands like construction and renovation of risers, which would benefit not only the land on which they are located but also other lands belonging to other farmers, these should be treated as items of benefit to the community and financed to the extent of 100 per cent by the State; and

(iii) the existing practice of financing the soil conservation programme on community land on 100 per cent basis should continue.

It is necessary to develop production programmes which, whilst preserving the soil, will gradually improve the economic condition of the people of the area. The strategy outlined below has been developed after careful consideration of the pros and cons of the matter;

(i) reclaiming land, where necessary, and providing minor irrigation, where possible, so as to encourage settled cultivation in villages and on terraced slopes on the lines adopted by the Savaras of Orissa and the Angamis in Nagaland, assuring at the same time inputs, fair price shops, communication and marketing facilities;

(ii) identifying the areas suitable for plantation crops such as tea, coffee, rubber, which would give subsidiary occupation for a family on the basis of the one hectare of plantation each;

(iii) developing gross reserves to support a subsidiary programme of animal husbandry; and

(iv) developing suitable areas for agro-silvicultural operations and commercial forest plantations, which would give full occupation to large number of village population.

An important change which has to be brought about is the change over from the hoe cultivation to plough cultivation and settled agriculture. For the resettled cultivation, the farmer has to learn the use of the plough, breeding of cattle and care of crops. Therefore, wherever such resettlement is being done, there has to be close supervision and training of the farmer families so that they can learn the new technology.

People living in the catchment areas necessarily require pasture lands for grazing and also land to raise their food requirements. Unless, therefore, an integrated watershed area development programme becomes a part of the treatment of catchment of the river valley project, the problem is not going to be solved. The first priority has to be provided for the needs of the local people for grazing, fuel wood and other requirements.

It is recommended that the maintenance of the assets created in the catchment should become an integral part of the whole scheme. The Committee would also recommend that top most priority should be given to the critically eroded areas of the catchments, particularly those lying within the 'hill areas'.

As scientific implementation of the soil conservation programme would require a multi-disciplinary programme and approach, the integrated watershed management approach cannot be planned or implemented unless a multi-disciplinary organisation is created.

It has also been found that the organisations handling soil conservation programmes are mono-disciplinary. Sufficient sub-divisions should be created which should be of multi-disciplinary type.

It would be necessary to provide refresher and reorientation courses to the officers of other departments involved in the problem of watershed management.

It is necessary to devise a cadre of soil conservationists with the basic background such as forest, soil science or science or agricultural engineering. It is equally necessary to acquaint the policy makers, administrators and financiers with the role of soil conservation in the areas of national and regional priori ties affecting various development programmes of the country.

Forests

Hill areas which are thickly forested or are in close prominity of thickly forested areas are highly suitable to horticulture, plantation crops, orchards, inter-culture, floriculture and pasture development.

Himalayan hills are suitable for raising mulbery trees and boost the silk production. Floriculture is turning out to be a flourishing business. The demand for cut flowers is increasing rapidly and hill areas are most suited to growing flowers of different kinds.

Hill areas are equally suitable for pasture development which will not only make a firm base for a Dairy development but would also check the growing menace of soil erosion.

Hill areas offer ample scope for agro-forestry based industries and handicrafts. In those of the hill areas which are richly endowed with forest wealth, there is enormous scope for setting up a large number of small industries. Those may include pulp paper and hard board, rosin and turpentine, sports goods, pencil industry, wickerwork, match industry, drug and pharmaceuticals, aroma chemicals and a large variety of other industrial enterprises.

The Committee recommends the following Action-oriented Programmes:

(i) Survey of land use and delineation of micro-watersheds requiring various degrees and types of treatment required and preparation of action plans.

(ii) Taking up required soil and water conservation measures including engineering as well as vegetative methods such as land shaping,, gull plugging, construction of check dams, soil binding plant species, etc. These have been covered in this book dealing with integrated watershed management.

(iii) Improvement of pastures and afforestation of degraded lands or waste lands belonging to Government or community should be taken up under social forestry and other programmes.

(iv) Introduction of species suitable for use as fodder, fuel and small timber of a very short duration. There should be adequate research support and the species should be chosen with care so that the local population can feel the impact of the benefits at the shortest possible time.

(v) Association of voluntary bodies and panchayats in a forestation and pasture development must be secured. Raising of community forests even on Government lands where ownership belongs to Government but the management given to the community where a strong panchayat or a voluntary body exists should be taken up.

(vi) Replacement of contractors by departmental felling or by forest labour cooperatives' and destruction of felling in vulnerable areas should be forthwith done. The Government of U.P. has issued an order that no felling should be carried in slopes above 45°, felling between 30° and 45° should be on selection system and no felling above 3000 meter altitude should be done. Some such regulations are necessary in other States also.

(vii) Animal Husbandry improvement scheme must be introduced simultaneously to reduce the number, while improving the quality of the stock. A very hard and unpopular decision at the political level is required to be taken about calling of any unremunerative cattle as recommended by the National forest Commission on Agriculture.

(viii) Creation of some biosphere reserves and more national parks so as to cover at least 10% of the forest areas in the Himalayas with very careful management in the above surrounds of such reserves and parks.

(ix) Diversion of forest land for non-forestry uses in the Himalayas must be stopped immediately. All lands under agriculture should be properly shaped and terraced and water courses regulated to do the least damage.

(x) In the matter of road construction whether under Central or State agencies cost for taking up conservation measures to eliminate damage due to cutting up of the hill sides should be included in the project itself, and its execution should be integral part of the roads programme.

(xi) Improvement should be carried out in the alpine pastures and grazing conditions along migration routes and in the lower ranges use for grazing by migratory graziers in winter. For that purpose there must be socio-economic coupled with agrostrological studies, regarding migratory grazing, conditions of graziers, intensity of grazing, availability of alpine pastures, the ownership, pattern of the livestock which the migratory graziers take to alpine pastures etc.

(xii) Horticulture (Himachal Pradesh as an example) is becoming very popular in this region. Care will have to be taken that this important economy for the development of hill areas does not result in degradation of environment and deforestation due to increasing demand of boxes for packing cases.

Steps would also have to be taken to ensure that it is made obligatory on orchards to replant the forest areas from where trees have been felled to pack the fruits.

The more important steps to eliminate the de forestation to meet the demand of packing would be the introduction of alternative packing material like hardboard. Alternative material packing would have to be persuaded vigorously if the expanding demand of boxes for packing is to be met, without large scale deforestation.

There is some hostility between the Forest Department staff and the local population, and this distress can be removed only through proper extension efforts and much closer contact between the two, so that each others problems and limitations are better understood.

The continual over-exploitation of trees and shrubs for fuel will go on unless a successful policy of first halting the trend and then reversing the process can be thought of and implemented. This is the most important aspect of the forest in the hill areas. Over exploitation of grass lands by the growing cattle and sheep population, without any attempt to regenerate the cover and control the grazing, has now led to a position where very little grass is available annually for the cattle and sheep wealth. The policy will have to aim at gradually halting this process of deterioration and building up the pastures of the hill areas so that a balance may be struck between the cattle and sheep wealth and available fodder in the first instance.

As a general policy, the nation has laid down that the purchase of minor forest produce in the forest areas must be departmentally managed so that the collectors of minor forest produce are given a fair price for their labourers. This policy has yet to be developed on large scale and it is also noticed that some States have a habit of going back on the national policy deliberately. A consistent policy will have to be developed to ensure that the tree wealth which gives the minor forest produce is not only maintained but developed so that steady source of income will be available to the local population. Thereby encouraging them to see to the preservation of these trees. Secondly, a fair price should be given to the produce not necessarily based on the labour charges alone but by complete departmentalisation of minor forest produce collection so that the urge to over-exploits is also curbed.

If the backward area is to be developed the local population should get every opportunity to benefit by the various developmental and employment opportunities that can arise in the area. Forests being an important part of the economy in the hill areas, every opportunity available in the forest for benefiting the local population and for giving them employment opportunities should obviously be availed of.

It will be useful to compile a list of purely protective forests for each of the States or Union Territories. It is obvious that very stringent tests will have to be applied to take out an area entirely from the orbit of productive forests and to earmark it for protection of terrain only.

The Committee advises that an immediate examination should be made of all such areas in the backward hill areas which should legitimately be brought within the classification of protected forests and steps taken to declare them as protected forests. Then a systematic programme should be taken up to see that the forest cover is brought back in all these protection forests within a reasonable time frame utilising the finances for 'social forestry'. This, in the view of the Committee, is of priority in the Himalayan hillranges.

Although the position may not have such serious consequence in the forest areas of the backward hill areas of the Deccan, still it is serious enough to war rant a similar action to identify the areas which should be put under protection forests. A period bound programme should then be taken up to ensure that such areas are afforested quickly and maintained as protection forests.

There are large forest areas which need replantation with economic species and commercial species. In the backward hill areas, economic forestry should include the types of timber trees and fodder trees which can solve the problem of raw material for village industries and fodder for the animal wealth.

Serious attempt should be made by forest administrations in the backward hilly areas to involve the local population in the operations connected with the forest managements which are remunerative so that the local population can see the benefit of the forests from their economic angle. Similarly, the involvement of the local population in collection of minor forest produce and paying them a fair remuneration for the collected produce is another aspect which needs wide acceptance by the forest departments and the Governments of the States.

In backward hill areas, therefore, where forests are important, the forest administration must learn to involve the local people so that they benefit from the forests and see the benefit in real terms.

Cold arid areas occur in the Ladakh region of Jammu and Kashmir (Ladakh & Kargil Districts) and Lahaul and Spiti region and Hongrang Valley of Himachal Pradesh. Land availability will be no limiting factor. The concerned States have already taken up a programme of afforestation, soil conservation and fodder development. This may be continued. Range management and grass land development should be an important part of the programme. The local people are not migratory and hence provision of improved fodder should have the utmost priority for the local livestock. Natural grass lands are not abundantly available in the region. The extremes of climate and high altitude also limit their choice For any large scale programme, pastures will have to be raised artificially where irrigation is available In summer, graziers from outside migrate to the area with their livestock and cause destruction of natural vegetation. It will be necessary to regulate the entry of such migratory graziers strictly in accordance with the carrying capacity of the grazing runs.

The Committee would strongly endorse the recommendation of the National Commission on Agriculture to set up a centre for forest research on cold desert to undertake forest research on the following:

(i) Identification of areas for pasture development and growing of forest trees to provide fodder for the grazing animals, small timber and fuel for the local population and stabilisation of areas threatened with erosion

(ii) Selection of suitable indigenous and exotic species and testing their performance.

(iii) Ecological and physiological studies in relation to photosynthesis and respiration at high elevation.

(iv) Studies of socio-economic aspects in relation to rangelands, soil and fater conservation.

(v) Study of the effect of shelter-belts and windbreak against high velocity wind.

Water Utilisation

The cost of construction of irrigation works in hill areas is higher in view of the topographical conditions. Following would appear to be the most appropriate methods for developing irrigational facilities in the hill areas:

(a) contour channels, with storages or diversion structures;

(b) lift irrigation from storages or running streams; and

(c) hydraulic rams.

In order to minimise the damage to contour channels, it is desirable to explore the possibilities of utilising polythene and alkathene pipes.

It has also been brought to the notice of the Committee that in Himachal Pradesh the rights of the farmers are standing in the way of optimum utilisation of kulh irrigation. We recommend that the State Government should review the entire position and devise suitable measures to ensure that there is full utilisation of the water and all the farmers with in the command of the kulh receive a share.

Lift irrigation from the running streams can go a long way in meeting the water needs for irrigation and drinking purposes. The Committee would recommend that valleys need to be identified where lift irrigation schemes are feasible and these should be suitably brought into a Master Plan of Construction quickly.

Fast rising prices of diesel oil and other fossil fuel have made it imperative to find out alternative sources of energy. This would point towards the need for further efforts in developing the technology for utilising the kinetic energy of water which mostly go waste at present and devices like Hvdraulic Rams, Himalayan Mill and Floating Mill which do not involve very high cost or long gestation period will be very useful.

Minor and Medium Irrigation in the Deccan Plateau are much simpler schemes than in the Himalayan Ranees. A Master Plan for Minor Irrigation in the Backward Hill areas is very necessary and the Committee recommends an early examination of the potential and a time bound programme for execution.

Land Use and Cropping

Quick growing economic species preferably handy fruit bearing (like walnut, peanuts and fodder) trees may be planted in these areas, if considered possible, so as to keep the interests of the owners alive.

Cultivation on hill slopes should be discouraged as far as possible as this causes erosion. Even the high value crops like potatoes which cause erosion should not be encouraged on slopes: There should be long term policy for covering such areas under perennial crops. Degrees of slopes beyond which cultivation should not be allowed may be specified for different regions.

As far as possible, horticulture should be encouraged in combination with sod cultivation. Sod should preferably be comprised of grass clover mixtures suitable to the locality. This will provide a very good combination for horticulture and animal husbandry both of which are complementary.

The Committee recommends strongly that the Agricultural Universities in the States which have the hill areas should now take up in right earnest firstly, applied research in the different zones of the various high yielding materials that have been developed in the country. At the same time, breeding of varieties suitable to the climatic conditions of the various zones of hill areas will have to be taken up strongly by these universities. The Committee recommends that the ICAR may not take active leadership in ensuring that this basic work for hill area development is now undertaken on a systematic manner.

The research has to be two-fold—firstly to identify the crops and zones where basic seed material of high quality can be supplied to the rest of the country and secondly pest and environmental problems leading to crop damage. Special attention will have to be paid in regional stations of the

Agricultural Universities to these problems. The land use in the hill areas as in other parts of the country has been guided mostly by the pressure of the population on land and the tendency is to bring under cereal production even marginal lands unsuitable for cultivation. This problem specially is acute in the hill areas leading thereby to cultivation on slopes and on this soil profiles thereby leading to very rapid soil erosion and a permanent loss to the productive areas. It is, therefore, necessary to see that the land use is adjusted to the potential of the land without leading to land deterioration. Pasture and horticultural development, the former on thin soil and the latter on sloping terrain which even now are largely practised in the hill areas, will have to replace the agricultural crops in marginal terrain. This can only be done by an active and intelligent extension approach.

Under the climatic conditions a horticultural or a vegetable crop can be much more profitable per unit of land than cereal production. The Agricultural Universities will have to take up detailed investigations of profitable land use under the climatic conditions and the right type of crop to be grown in the various zones under various environmental conditions. This will be an exercise extending the idea that has been postulated by the N'CA about fitting the right type of food crop to the hill area. The Committee recommends that this research work should be done.

Today no householder in the backward areas with the paucity of an administrative support and communications will risk the possibility or not producing a certain minimum food crop for his own family and depend on other people for his food. The change ever, therefore, cannot take place all over the lull areas and on mere statement of policy but some selective approach will be necessary. The problem can be solved if wherever the good crops are replaced by new cash crops whether cereal, agricultural raw materials, horticulture or animal husbandry products, suit able marketing facilities are built into the system to buy the production at their fair prices and take them off the hands of the farmers of the area. Parallel to this organisation there has to be an organisation to supply the food crop prevalent in the area in fair price shops so that the farmer can buy his necessities from the fair price shops.

An area approach of development taking the more productive zones which can be still more productive and on the communication routes will have to be taken up first. The administrative support and the technical support will have to be laid on as a necessary part of the programme. The Committee recommends that this vast approach of development and change over of the cropping pattern to the best use should be brought about in the hill areas at least substantially by 2000 A.D. A vast programme should be built into the Five Year Plans.

It is necessary that the productivity and income per unit of land is increased so that the hill areas can sell in the plains and get their needs from the plains balancing or improving the balance towards their side by the greater income per unit of land.

An important problem in expanding horticulture in the hill areas is the difficulty of communications in the Himalayan areas making it difficult to transport perishable fruits quickly to the consurning markets. Firstly, it is necessary to ensure that where-ever active propagation of horticulture is being done as a remunerative replacement for the subsistence cereal economy; care should be taken to see that a suitable marketing system can be run in these areas to remove the produce quickly to the marketing areas. Where communications are difficult or cannot be established before the trees come into fruition, other methods of handling the fruits so as to give a remunerative return to the farmer needs examination.

Achievement of high productivity is not sufficient by itself. The grower must get incentive return so as 10 enable him to maintain the orchard properly. This is only possible if the post-harvest handling and marketing systems are modernised and streamlined.

Presently, all packing of fruits in the hills is done in wooden cases. The availability of wood will outstrip the demand. Therefore, the forests will be devasated and then Horticulture Industry will reach a dead end by the end of the century. Introduction of alternative packing cases is, therefore, a must.

The Committee recommends that firstly research work in fruit processing and preservation should be carried out as a priority subject in the Agricultural Universities in the States where hill areas are important. Secondly, imaginative schemes for fruit preservation should be launched in time to take off all the fresh fruit in difficult areas and the fresh fruit during flush seasons in the regions where horticulture is being deliberately expanded.

It is proposed that some minimum limit say 15 per cent for use of fruit juices in all sorts of drinks may be recommended. The incidence of duties and taxes should be reduced with the increase in percentage of juice in such drinks.

Horticulture should be treated at par with the industries in the backward areas for the purposes of advance of capital and interest.

Horticultural planning should be integrated covering all important aspects such as Production (Production of planting material, orchard management etc.) post-harvest handling, marketing and processing. All these should have a strong backing of Research.

Temperate vegetables can be grown best in the middle levels of the hill areas both in the Himalayan ranges and in the Western Ghats. Proper planning of the right type of vegetables for the various areas and the adjustment of the seasons to give them a scarcity value in the plains market in various seasons should now be taken up seriously by the Horticulture Administration of States. Along with the increase of irrigation, providing for a tree crop routine in parts of the hill areas, introduction of vegetable growing in the crop routines will give greater remuneration to the farmers than following purely cereal routine.

Plantations have not come up in a big way in the Himalayan hills in the Western and Middle parts. Some amount of research will have to be done about the possibility of plantation crops in these areas. The country has found new areas for planting tea, rubber, coffee and so on through the various Boards taking an active part on promotion and identifying suitable areas with suitable climatic conditions. This work should be done in a wider way in the Himalayan hills. In all the hill areas, the relevant varieties of trees will have to be identified for tasar programme and for a fodder tree programme. As pasture and animal husbandry development is of special significance in both the Himalayan hills and the Western Ghats, a fodder tree plantation will be of special benefit to the rural population.

The hill areas provide good scope for floriculture. The need is to organise and expand production and marketing to make floriculture more lucrative.

Another direction in which the hill areas can specialise is aromatic and medicinal plants. Many of those are wild growth in the hill forests.

The Committee recommends that the Central Silk Board should explore the possibility of expanding sericulture extensively in the hill areas. Ex tension of tasar culture throughout the oak belt of the Himalayas should be vigorously pursued by the Central Silk Board.

The Central Silk Board should organise seed multiplication and distribution, if necessary, by importing exotic strains. The Department of Agriculture can deal with sericulture aspect up to the processing stage of cocoons. Short terms in service training to junior staff and familiarisation training of farmers and rearers should be introduced.

Possibility also exists for taking up tasar culture, which is now largely confined to Assam, through out the oak belt of the Himalayas. The development of tasar hybrid, which thrives well on oak, has production of high quality material. Similarly muga culture at present practised in Assam can be extended to other parts of the north eastern regions.

Apiculture, *i.e.* production, collection and marketing of honey and honey products could be a useful subsidiary occupation giving supplemental in come to the people in the hills. There is great scope for increasing honey yields through organised apiculture, horticulture and forests. Detailed survey of the vegetation of forests with regard to floristic composition should be carried out.

Lack of Pollinisers

A few years ago the farmers had many apple varieties: Commercial, Golden, Jonathan, Red Gold, Kali Devi, Red and Royal Delicious. These provided compatible pollen to other varieties and good fruit set. Since 1980 because of the better market value of Royal Delicious, farmers have uprooted other varieties and planted Royal Delicious on a large scale. All new orchards have been planted only with Royal Delicious. However, Royal Delicious is completely self-sterile and requires pollen from other compatible or polliniser varieties for fruit set to occur. Appropriate ratios of apple pollinisers do not exist and some farmers have no polliniser trees.

Farmers are now aware of the need for pollinisers, for good yields and apple quality, and are planting pollinisers. Some farmers are grafting pollinisers on to Royal Delicious trees. Farmers in Shimla District have grafted a polliniser variety called *Snowdrift*, selected from crab apple, a wild apple variety. Snowdrift acts as the bouquet and has a long flowering period of more than a month, providing pollen to many other varieties, whether early or late blooming.

The main polliniser varieties now planted in Kullu District include Kali Devi, Commercial and Golden. In Shimla District the most commonly used polliniser varieties are Golden, Red Gold, Tydemann and Early Worcester.

Short-term Solution

Farmers place polliniser branches in plastic bags filled with water and hang these on Royal Delicious trees. This type of pollination management is called *'Bouquet Pollination'*, and the polliniser branches *'Bouquets'*.

Bouquet pollination was first started by Mr. Gulab Singh in 1994 in Kullu District. Previously fruit set in his orchard had been down to only 10%, but using bouquet pollination he obtains excellent fruit set. All the farmers in nearby villages went to Mr. Gulab Singh's orchard to investigate his pollination management. They were very impressed and the following year all used bouquet pollination as a short-term solution for polliniser management. It increased fruit yield by about 50%.

Unexpected Problems

1. Some farmers who do not have enough polliniser trees are stealing polliniser branches from other orchards. There are many cases of such theft. As a result whenever farmers see somebody with scissors, they chase the intruder away. We met one farmer in the Manali Valley carrying a load of blooming polliniser branches. We asked him to stop as we wanted to photograph him with these branches. He was afraid and told us that he brought these branches from his own orchard and was going to hang them in his uncle's orchard.
2. The large-scale use of plastic bags for hanging pollinisers in trees has led to a four-fold increase in the price of plastic bags!

Lack of Pollinators

Lack of pollinating insects due to the heavy use of insecticides over the past decades is another factor. Farmers have been spraying pesticides eight to ten times each season including the blooming period of apple trees. Bees, butterflies, and moths on apple flowers have all disappeared.

Hiring Honeybees for Pollination

Insect pollinators are still needed to transfer the pollen grains from the polliniser to the main variety. For this purpose farmers are renting honeybees (*Apis cerana* and *Apis mellifera*). The Department of Horticulture rents *Apis mellifera* colonies at the rate of Rs 250 per colony, (Rs 200 as security and Rs 50 as rent). Private beekeepers charge Rs 800 (Rs 500 as security, and Rs 300 as rent).

Honeybees for Rent

The number of bee colonies available for rent from the Department of Horticulture is low and the demand for bees is very high. Therefore, to provide bee colonies for a number of farmers, very weak colonies are offered, with only one or two frames covered with bees.

Some commercial beekeepers migrate their bees to the Indian plains during winter until June. They are not interested to bring their bees back earlier for apple pollination because they obtain abundant honey from berseem and sunflower in the plains. These crops bloom for long periods and produce much more nectar than apple. The beekeepers earn much better money from the sale of honey, than by renting bees for apple pollination. Now these beekeepers are being encouraged to rent their bees for pollination by increased rental fees.

Farmers are ready to pay any price for hiring bees: some farmers hire *Apis cerana* in log hives at the rate of Rs 300 per colony. The State Government is making efforts to encourage farmers to keep their own bees for pollination and has created the Beekeeping Development Office (BKDO), within the Department of Horticulture. BKDO sells an *Apis mellifera* bee colony at the highly subsidised rate of just Rs 300: the market price is over Rs 2000.

Dr. Y.S. Parmar University of Horticulture and Forestry also provides bee colonies free of charge to apple farmers and offers training on how to use bees for pollination.

In Kullu District, there is only one commercial beekeeping farm, Yama Bee Farms, renting *Apis mellifera* for pollination. In Shimla there are more commercial beekeepers who rent out bees for apple pollination.

Need for Training Apple Growers to Manage Bees

Farmers have a great awareness of the role of honeybees in pollination, but they have no knowledge of bee behaviour. They want to keep bees at any cost: a farmer stole a honeybee colony from a neighbour one night and kept it in his orchard in the hope that the bees would pollinate his apples during the night. Early in the morning, he put the bee colony back in the owner's orchard!

Use of Pesticides

Farmers would spray chemicals even if there is no disease to prevent the outbreak of apple scab and red apple mites. They are now aware of the harmful impact of insecticides on natural pollinators, and have reduced pesticide application from ten, to five or six sprays a year, usually using the least toxic chemicals. They spray pesticides when insect pollinators are not present-a week before or after apple flowering. Some large-scale farmers are controlling apple pests through biological pest control-rearing various predator insects against the pests.

Hand Pollination

Some farmers in Shimla District are very progressive and are hand pollinating their apples to ensure the flowers are adequately pollinated. According to them this is most important to ensure the quality and yield of apple. They can afford to hire labour for hand pollination of their orchards. Anthers are picked from the flower of the polliniser variety at the 'balloon' stage (partially open flowers) and dried at 22°C overnight to release the pollen grains. These are mixed up with a little white flour or skimmed milk powder and applied to flowers of the main variety (within two days of flowers opening) with the help of a brush. Between 20 and 50 flowers are hand pollinated on one branch and then honeybees transfer this pollen to other flowers of the branch.

The role of honeybees is crucial in apple pollination, both in hand pollination of the crop and polliniser management, either by grafting, or hanging pollinisers on the main varieties.

Climate Change

Climate change during the past eight years has played a critical role in apple pollination failure. There are rains during the flowering season which affect pollination by wind and insects. Low temperatures also adversely affect fruit set in apple.

Temperature Management

If there is a possibility of frost the farmers collect grasses and burn them in the orchards to raise the temperature by a few degrees. Some farmers in Shimla District spray the orchard with pure water at midnight to avoid frost the following morning. Farmers also spray chemicals such as borax, TSO or some alternative insecticides on a limited scale to delay flowering by about one week to prevent flowering when the temperature is very low. To date the farmers have no remedy to prevent rain during apple flowering!

Conclusion

Apple farmers in Himachal Pradesh are now well aware of the need for pollination and are making all possible efforts to ensure pollination occurs: managing pollinisers and pollinators, and opting for strategies to avoid the effects of low temperatures or frost on pollination and fruit set.

Government institutions in the State have an explicit mandate to educate the apple farmers on how to achieve good yields, but they have limited resources to cope with the large-scale extension needed in this area.

Technologies for Water Harvesting and Soil Moisture Conservation

Water is essential for all life and is used in many different ways-for food production, drinking and domestic uses and industrial use. It is also part of the larger ecosystem on which bio diversity depends. Precipitation, converted to soil and ground-water and thus accessible to vegetation and people, is the dominant precondition for bio-mass production and social development in drylands. The amount of available water is equivalent to the water moving through the landscape. It also fluctuates between the wet and dry periods. Fresh water scarcity is not limited to the arid climatic regions only. Even in areas with good supply, the access to safe water is becoming a critical problem. Lack of water is caused by low water storage capacity, low infiltration capacity, large inter-annual and annual fluctuations of precipitation and high evaporative demand.

Irrigation in Africa is a privilege since, the cost of providing irrigation to one hectare could be as high as US$ 20 000/ha and therefore cannot be implemented. Even for small-scale irrigation using ground-water, well (both open and tube well) construction is costly. It may amount to more than US$ 10 000 to provide irrigation water for one hectare of land.

Irrigation potential which exists in South and East African countries is much more than the presently irrigated area. The geographical area, irrigated area at present and irrigation potential for selected countries. The table indicates the possibilities of bringing a larger area under irrigation in these countries for food security. A variety of essential soil moisture and water conservation technologies must be adopted to reduce the cost of irrigation, extend it throughout and promote sustainable small-scale irrigation on a watershed basis. These technologies are essential especially in drought-prone areas. Even though drought is a purely natural calamity caused by the failure of (monsoon) rain, it can be minimized by careful planning and operation. During good rainy years, excess rainwater should be stored in the soil and also underground using suitable soil moisture conservation measures and water harvesting structures on a watershed basis. This stored water can subsequently be used for irrigation.

Conceptual Approach

Watershed development and management implies an integration of technologies within the natural boundary of a drainage area for optimum development of land, water and plant resources, to meet the people's basic needs in a sustained manner. A watershed is an area from which runoff resulting from precipitation flows past a single point into a large stream, river, lake or pond. Each watershed is an independent hydrological unit. It has become an acceptable unit of planning for optimum use and conservation of soil and water resources.

Area, population, irrigation details of selected countries

Details	Malawi	Tanzania	Zambia	Zimbabwe	Mozambique
Area (M ha)	11.85	94.50	75.26	39.08	80.18
Population (Million)	10.08	28.85	9.20	11.00	15.50
Rainfall (mm)	1014	937	1011	652	969
Cultivable area (M ha)	3.6	40	16.35	--	36.0
Cultivated area (M ha)	2.1	6.3	1.03	2.80	3.6
Irrigation Potential (M ha)	0.162	0.83	1.40	0.55	3.30
Irrigated area (M ha)	0.09	0.15	0.15	0.14	0.11
Population per sq km	92	31	12	28	19

The concept of integrated watershed development refers to the development and management of the resources in the watershed to achieve higher sustainable production without deterioration in the resource base and any ecological imbalances. This concept requires the formulation and implementation of a package of programmes with activities for optimum resource use in the watershed without adversely affecting the soil and water base or life supporting system. The concept assumes more importance in the context of planning for sustained development. Watershed development aims at preventing watershed degradation resulting from the interaction of physiographic features. It eliminates unscientific land use, inappropriate cropping patterns and soil erosion, thereby improving and sustaining productivity of resources leading to higher income and living standards for the inhabitants in the watershed area. It therefore involves restoration of the ecosystem, protecting and utilizing the locally available resources within a watershed to achieve sustainable development.

Rainfall failure occurs once every 3 to 5 years and is usually below 50% of the average annual rainfall of the region. During periods of rainfall failure, the ground-water level lowers since fluctuations in the water table levels depend on the rainfall when both surface and ground-water availability becomes critical. Drought begins to prevail and there is difficulty to cope up with the water demand during this period. Similarly, in some locations or areas water shortage is observed just before the rainy season commences. These two situations can be managed if suitable soil and moisture conservation measures are systematically implemented on a small watershed basis.

There are always strong links between soil conservation and water conservation measures. Many actions are directed primarily to one or the other, but most contain an element of both. Reduction of surface runoff can be achieved by constructing suitable structures or by changes in land management. Further, this reduction of surface runoff will increase infiltration and help in water conservation.

Appropriate Structures and Their Functions

To increase the period of water availability and overcome water scarcity in drought years, the following activities can be implemented in the field for a compact, viable watershed of about 200-500 ha.

Soil and water conservation can be approached through agronomic and engineering procedures. Agronomic measures include contour farming, off season tillage, deep tillage, mulching and providing vegetative barriers on the contour. These measures mainly prevent soil erosion but will also help in improving soil moisture availability in the watershed.

The engineering measures adopted differ with location, slope of the land, soil type, amount and intensity of rainfall. Depending on these parameters, the methods commonly used are contour trenching, contour stone walls, construction of temporary and permanent check dams and gully plugging structures. Additionally, percolation ponds, silt detention tanks and irrigation tanks are constructed to harvest water and recharge it to the ground-water for use in agriculture (irrigation). Farm ponds can also be constructed for every 4-5 ha in the watershed to provide protective/supplemental irrigation. The above soil and water conservation management and water harvesting programme should be implemented in an integrated manner on a catchment/watershed basis.

Functions of the Structures

Contour bunds, contour barriers (vegetative and stone), contour trenches and contour stone walls will not only prevent soil erosion but also obstruct the flow of runoff water. Consequently, the obstructed water will increase the soil moisture and recharge the ground-water in the area.

Check Dams: This may be a temporary structure constructed with locally available materials. The various types are: Brush wood dam, loose rock dam and woven wire dam. The main function of the check dam is to impede the soil and water removed from the watershed. This structure is cheap, but lasts about 2-5 years. The cost of the structure is about US$ 200-400 depending on the materials used, the size of the gully and the height of the obstruction (dam). A permanent check dam can be constructed using stones, bricks and cement. Small earth work is also needed on both sides. Costs vary from US$ 1 000-3 000 depending upon the length and height of the dam. A little water is also stored above the dam. This water recharges the ground-water.

Percolation Pond: The percolation pond is a multipurpose conservation structure depending on its location and size. It stores water for livestock and recharges the ground-water. It is constructed by excavating a depression, forming a small reservoir or by constructing an embankment in a natural ravine or gully to form an impounded type of reservoir. The cost of this type of structure is estimated at around US$ 5 000-10 000. The capacity of these ponds or tanks varies from 0.3 to 0.5 mcft (10 000-15 000 m^3). Normally 2 or 3 fillings are expected in a year (season) and hence the amount of water available in one year in such a tank is about 1 mcft to 1.5 mcft (30000-45000 m^3). The cost for one mcft (30 000 m^3) varies from US$5 000 to 15000. This quantity of water, if it is used for irrigation, is sufficient to irrigate 4-6 hectares of irrigated dry crops (maize, cotton, pulse, etc.) and 2-3 hectares of paddy crop.

Irrigation Tank: The main function of this storage structure is irrigating crops. It is constructed below the above-mentioned structures in a watershed. In Tamil Nadu, India, each tank irrigates from 10 to 5 000 hectares. In south India, there are about 2, 000, 000 tanks, irrigating about 3.5 million hectares. Earthen bunds are reinforced with masonry to collect and store rainwater for irrigation. The cost of this tank (dam) depends upon the size, location and site condition. It costs about US$ 1 500-2 500 to irrigate 1 ha of land. Water from the tanks is normally used to grow paddy crop.

Apart from the above, to increase moisture availability to agricultural and tree crops, in situ moisture conservation techniques must be adopted in addition to the large scale soil and moisture conservation and water harvesting structures in the watershed. The following are some of the *in situ* moisture conservation measures which can be practised in the watershed to increase production.

For agricultural crops, the measures adopted are forming ridges and furrows, broad bed and furrows, basins, tie ridging (random tie ridges) and water spreading.

For tree crops micro catchment, saucer basin, semi-circular bund, crescent shaped bunds, V ditch technology, catch pits and deep pitting can be practised.

In addition to the above measures and structures, small storage structures with a water storage capacity for an area of about 0.4 to 0.5 ha can be constructed in large numbers one for every 10 to 20 ha catchment or watershed at the foot hills slopes and hilly areas. These storage facilities would attenuate the floods during storms. These measures will also ensure soil moisture for good growth of trees grown down stream recharging the ground-water in the region and making available more water for drinking and irrigation water.

Construction Procedures

The procedure for the construction of tanks and check dams, on a watershed basis, is as follows. Detailed studies are first made of the watershed to determine its current erosion condition, land use and water balance condition. Based on these data, decisions are made on the construction of check dams (temporary/permanent), percolation tanks and irrigation tanks. The community in the watershed should be involved in planning and selecting the type of structures and locations of the structures. If a large quantity of water is lost through runoff from the watershed, 2-3 percolation tanks or one irrigation tank can be planned based on water availability. A number of check dams can be provided to arrest erosion and to plug the gullies in the rivulets at appropriate places.

The construction can be done using human labour if necessary or by a mix of mechanical and human effort. In India, these works are mostly executed by labourers to give employment to the village people. Only 10% of the work is done mechanically (bulldozer) *i.e.*, digging the foundation and excavating earth to form bunds. Otherwise 80-90% of the work can be done by machines, if employment is not a criterion.

The cost of construction of the above-mentioned structures depends upon the soil type, the size of the tank and season of construction. The proportions of earth and masonry works for the three different structures are as follows.

Type of structures	Earthwork (%)	Masonry and concrete
Check dam	10	90
Percolation tanks and irrigation tanks	40-70	30-60

The average costs of construction of percolation ponds are as follows: (actuals taken from the estimates).

Type of construction	Labour	Capacity of the tank (mcft)	Earthwork cost in US$	Material cost in US$	Cost/Unit (1 mcft or 30 000 m³)
Percolation pond 1	90 %	0.43	4 000	300	10 000
Percolation pond 2	93 %	0.43	4 200	300	10 000
Percolation pond 3	70 %	0.30	2 000	4 500	25 000
Percolation pond 4	77 %	0.25	2 000	5 000	28 000
Percolation pond 5	75 %	0.25	2 000	3 000	25 000

The actual construction is undertaken when the community is free from agricultural work and the works are spaced in such a manner that the construction will be completed before the rain (monsoon) commences, so that the structure will be used immediately. Small structures are completed within 2-3 months while a large one takes about 7-9 months.

These tanks or dams are constructed by Government without the involvement of people during construction. The maintenance is normally not done but is the responsibility of the Government. If the people are involved in constructing the structure they feel ownership and will take responsibility for maintenance. Since the work is given to the contractors the quality of the construction is also questionable. Wherever the community is involved in the construction, they undertake the maintenance of the structure themselves.

Users' (farmers') Participation

The efficiency of a tank or pond depends on the organization of its beneficiaries. This organization of beneficiaries are required for each and every tank, especially for maintenance and rehabilitation. The role of the water users' association (WUA) or farmers' organization (FO) in general is as follows.

- Planning including selection of site.
- Construction.
- Distribution of water in irrigation tanks.
- Maintenance of the structures.
- Mobilization of resources for the work, management and maintenance.
- Involving themselves in the work.
- Resolution of the conflicts.

Linkages should be established between the farmers and various organizations. The NGOs are required as a catalyst and to create awareness among the farmers about efficient utilization of the resources.

It is said that users' participation is necessary for better utilization and management of the resources. In practice this participation is very difficult to achieve. To some extent farmers' participation in the management and utilization of irrigation tanks is forthcoming. However, involving farmers is not such an easy task with check dams and percolation ponds, since there is no direct benefit to the farmers as in the case of irrigation tank.

Involvement of farmers in system management and project management from planning to operation and management of the system should be stressed. Coordination among the association members has to be intensified for proper guidance.

3

Planning Concept, Design and Construction

When selecting the construction site of check dams, percolation ponds and tanks, the following criteria can be considered.

Check Dams: This is constructed across rivulets and gullies to control erosion, prevent gully formation and to arrest the flow of water to allow it to go underground. A number of such obstructions will be useful for soil and moisture conservation measures. Inexpensive, temporary structures can be constructed using vegetation, stone or brushwood, available at the site. Large numbers can be provided to reduce erosion and formation of gullies. Permanent check dams can be located at the junction of one or two streams or gullies using masonry structures.

Percolation Ponds: The following factors may be considered:

- It should not be located in heavy soils or soils with impervious strata, otherwise the top soil should be porous.
- Suitable and adequate soil should be available for forming embankments.
- The ideal location of the pond will be on a narrow stream with high ground on either side of the stream.
- Simple, economic and efficient surplus arrangement should be possible.

Pond size should be decided on the basis of the catchment area and the number of fillings possible for the pond in the area.

Irrigation Tanks: The location should be such that it should receive water from a large catchment area.

- There should be land below the site suitable for irrigation.
- Cooperation and coordination of the farmers and the community is essential to use the water.
- The location should be such that it will be a narrow point with high ground and wide open space in front of the tank location so that a large quantity can be stored with minimum cost.

After the plan and estimates are prepared and approved by the appropriate authority, construction can be entrusted to village people (beneficiaries) in case of check dams or percolation tanks and to the WUA and/or FO in case of irrigation tanks. Not much operation and management is needed in the case of check dams and percolation tanks but irrigation tanks must be visited regularly to release water and conduct repairs and maintenance. The WUA or FO can be involved from the time of construction, operation of the system and maintenance of the structure.

Performance Assessment and Lessons Learned

Farmers should always be involved in planning and executing field programmes. Management and new techniques should not become a continuous burden to the farmers. The design and implementation of a conservation and rehabilitation programme is likely to be most successful, if they are the product of interdisciplinary work between technical staff and those experienced in economic and social issues.

Based on the above background, the author had an opportunity to study many projects implemented by NGOs, government departments and farmers on water harvesting and soil moisture conservation in small watersheds in India. After visiting all the projects and following discussions with the people concerned, the following assessments were made about their projects and the lessons learned from the experiences.

In one village, owing to the inspiration of a dedicated NGO, the village has dramatically improved after the villagers executed all the development work including soil conservation and water harvesting works. Though the annual rainfall was only 450-500 mm, the ground-water table was maintained at a depth of 6-7 m and there is no water scarcity even during drought years in that village. The farmers are using drip irrigation for widely spaced fruit and orchard crops and are planting trees in barren and common lands.

In another large project (40 000 ha) funded by the World Bank, soil and water conservation and water harvesting measures were conducted on a watershed basis. Though it was a Government project it was implemented with the help and involvement of the community from the planning stage to execution. Consequently the project achieved its goal. The farmers obtained 150-200% more yield. The ground-water table in the watershed was raised and the community had no difficulty to get water. At the same time, another project failed and expected results were not been achieved because farmers were not involved. Therefore it is clear that water availability can be increased if soil conservation and water harvesting measures are implemented on a watershed basis involving the community. In addition more water will be available for small-scale irrigation, which will assist in improving the people's living conditions and food security in the coming years.

Bael Fruit

Aegle Marmelos Correa

Though more prized for its medicinal virtues than its edible quality, this interesting member of the family Rutaceae is, nevertheless, of sufficient importance as an edible fruit to be included here. The bael fruit, *Aegle marmelos* Correa (syns. *Feronia pellucida* Roth., *Crataeva marmelos* L.), is also called Bengal quince, Indian quince, golden apple, holy fruit, stone apple, *bel, bela, sirphal, maredoo* and other dialectal names in India; *matum* and *mapin* in Thailand; *phneou* or *pnoi* in Cambodia; *bau nau* in Vietnam; *bilak,* or *maja pahit* in Malaya; *modjo* in Java; *oranger du Malabar* in French; *marmelos* in Portuguese. Sometimes it is called elephant apple, which causes confusion with a related fruit of that name, *Feronia limonia* Swingle (q.v.).

Description: The bael fruit tree is slow-growing, of medium size, up to 40 or 50 ft. (12-15 m) tall with short trunk, thick, soft, flaking bark, and spreading, sometimes spiny branches, the lower ones drooping. Young suckers bear many stiff, straight spines. A clear, gummy sap, resembling gum arabic, exudes from wounded branches and hangs down in long strands, becoming gradually solid. It is sweet at first taste and then irritating to the throat. The deciduous, alternate leaves, borne singly or in 2's or 3's, are composed of 3 to 5 oval, pointed, shallowly toothed leaflets, 1 1/2 to 4 in (4-10 cm) long, 3/4 to 2 in (2-5 cm) wide, the terminal one with a long petiole. New foliage is glossy and pinkish-maroon. Mature leaves emit a disagreeable odour when bruised. Fragrant flowers, in clusters of 4 to 7 along the young branchlets, have 4 recurved, fleshy petals, green outside, yellowish inside, and 50 or more greenish-yellow stamens. The fruit, round, pyriform, oval, or oblong, 2 to 8 in (5-20 cm) in diameter, may have a thin, hard, woody shell or a more or less soft rind, gray-green until the fruit is fully ripe, when it turns yellowish. It is dotted with aromatic, minute oil glands. Inside, there is a hard central core and 8 to 20 faintly defined triangular segments, with thin, dark-orange walls, filled with aromatic, pale-orange, pasty, sweet, resinous, more or less astringent, pulp. Embedded in the pulp are 10 to 15 seeds, flattened-oblong, about 3/8 in (1 cm) long, bearing woolly hairs and each enclosed in a sac of adhesive, transparent mucilage that solidifies on drying.

Origin and Distribution: The tree grows wild in dry forests on hills and plains of central and southern India and Burma, Pakistan and Bangladesh, also in mixed deciduous and dry dipterocarp forests of former French Indochina. Mention has been found in writings dating back to 800 B.C. It is cultivated throughout India, mainly in temple gardens, because of its status as a sacred tree; also in Ceylon and northern Malaya, the drier areas of Java, and to a limited extent on northern Luzon in the Philippine Islands where it first fruited in 1914. It is grown in some Egyptian gardens, and in Surinam and Trinidad. Seeds were sent from Lahore to Dr. Walter T. Swingle in 1909 (P.I. No. 24450). Specimens have been maintained in citrus collections in Florida and in agriculture research stations but the tree has never been grown for its fruit in this state except by Dr. David Fairchild at his home, the "Kampong", in Coconut Grove, after he acquired a taste for it, served with jaggery (palm sugar), in Ceylon.

Climate: The bael fruit tree is a subtropical species. In the Punjab, it grows up to an altitude of 4,000 ft. (1,200 m) where the temperature rises to 120° F (48.89° C) in the shade in summer and descends to 20° F (-6.67° C) in the winter, and prolonged droughts occur. It will not fruit where there is no long, dry season, as in southern Malaya.

Soil: The bael fruit is said to do best on rich, well-drained soil, but it has grown well and fruited on the oolitic limestone of southern Florida. According to L. B. Singh (1961), it "grows well in swampy, alkaline or stony soils"... "grows luxuriantly in the soils having pH range from 5 to 8". In India it has the reputation of thriving where other fruit trees cannot survive.

Varieties: One esteemed, large cultivar with thin rind and few seeds is known as 'Kaghzi'. Dr. L.B. Singh and co-workers at the Horticultural Research Institute, Saharanpur, India, surveyed bael fruit trees in Uttar Padesh, screened about 100 seedlings, selected as the most promising for commercial planting: 'Mitzapuri', 'Darogaji', 'Ojha', 'Rampuri', 'Azamati', 'Khamaria'. Rated the best was 'Mitzapuri', with very thin rind, breakable with slight pressure of the thumb, pulp of fine texture, free of gum, of excellent flavour, and containing few seeds.

Only the small, hard-shelled type is known in Florida and this has to be sawed open, cracked with a hammer, or flung forcefully against a rock. Fruits of this type are standard for medicinal uses rather than for consuming as normal food.

Propagation: The bael fruit is commonly grown from seed in nurseries and transplanted into the field. Seedlings show great variation in form, size, texture of rind, quantity and quality of pulp and number of seeds. The flavour ranges from disagreeable to pleasant. Therefore, superior types must be multiplied vegetatively. L.B. Singh achieved 80% to 95% success in 1954 when he budded 1-month-old shoots onto 2-year-old seedling bael rootstocks in the month of June. Experimental shield-budding onto related species of *Afraegle* and onto

Swinglea glutinosa Merr. has been successful. Occasionally, air-layers or root cuttings have been used for propagation.

Culture: The tree has no exacting cultural requirements, doing well with a minimum of fertilizer and irrigation. The spacing in orchards is 25 to 30 ft. (6-9 m) between trees. Seedlings begin to bear in 6 to 7 years, vegetatively propagated trees in 5 years. Full production is reached in 15 years. In India flowering occurs in April and May soon after the new leaves appear and the fruit ripens in 10 to 11 months from bloom–March to June of the following year.

Harvesting: Normally, the fruit is harvested when yellowish-green and kept for 8 days while it loses its green tint. Then the stem readily separates from the fruit. The fruits can be harvested in January (2 to 3 months before full maturity) and ripened artificially in 18 to 24 days by treatment with 1,000 to 1,500 ppm ethrel (2-chloroethane phosphonic acid) and storage at 86° F (30° C). Care is needed in harvesting and handling to avoid causing cracks in the rind.A tree may yield as many as 800 fruits in a season but an average crop is 150 to 200, or, in the better cultivars, up to 400.

Keeping Quality: Normally-harvested bael fruits can be held for 2 weeks at 86° F (30° C), 4 months at 48.2° F (9° C). Thereafter, mold is likely to develop at the stem-end and any crack in the rind.

Pests and Diseases: The bael fruit seems to be relatively free from pests and diseases except for the fungi causing deterioration in storage.

Food Uses: Bael fruits may be cut in half, or the soft types broken open, and the pulp, dressed with palm sugar, eaten for breakfast, as is a common practice in Indonesia. The pulp is often processed as nectar or "squash" (diluted nectar). A popular drink (called "sherbet" in India) is made by beating the seeded pulp together with milk and sugar. A beverage is also made by combining bael fruit pulp with that of tamarind. These drinks are consumed perhaps less as food or refreshment than for their medicinal effects.

Mature but still unripe fruits are made into jam, with the addition of citric acid. The pulp is also converted into marmalade or sirup, likewise for both food and therapeutic use, the marmalade being eaten at breakfast by those convalescing from diarrhea and dysentery. A firm jelly is made from the pulp alone, or, better still, combined with guava to modify the astringent flavour. The pulp is also pickled.

Bael pulp is steeped in water, strained, preserved with 350 ppm SO_2, blended with 30% sugar, then dehydrated for 15 hrs at 120° F (48.89° C) and pulverized.

The powder is enriched with 66 mg per 100 g ascorbic acid and can be stored for 3 months for use in making cold drinks ("squashes"). A confection, bael fruit toffee, is prepared by combining the pulp with sugar, glucose, skim milk powder and hydrogenated fat. Indian food technologists view the prospects for expanded bael fruit processing as highly promising.

The young leaves and shoots are eaten as a vegetable in Thailand and used to season food in Indonesia. They are said to reduce the appetite. An infusion of the flowers is a cooling drink.

Food Value Per 100 g of Edible Portion*

Water	54.96-61.5 g
Protein	1.8-2.62 g
Fat	0.2-0.39 g
Carbohydrates	28.11-31.8 g
Ash	1.04-1.7 g
Carotene	55 mg
Thiamine	0.13 mg
Riboflavin	1.19 mg
Niacin	1.1 mg
Ascorbic Acid	8-60 mg
Tartaric Acid	2.11 mg

*Fresh bael fruit, as analyzed in India and in the Philippines.

The pulp also contains a balsam-like substance, and 2-furocoumarins-psoralen and marmelosin ($C_{13}H_{12}O_3$), highest in the pulp of the large, cultivated forms.

There is as much as 9% tannin in the pulp of wild fruits, less in the cultivated types. The rind contains up to 20%. Tannin is also present in the leaves, as is skimmianine.

The essential oil of the leaves contains *d-limonene*, 56% *a*-d-phellandrene, cineol, citronellal, citral; 17% p-cyrnene, 5% cumin aldehyde. The leaves contain the alkaloids *O*-(3,3-dimethylallyl)-halfordinol, *N*-2-ethoxy-2-(4-methoxyphenyl) ethylcinnamide, *N*-2-methoxy-2-[4-(3',3'-dimethyalloxy) phenyll]ethylcinnamide, and *N*-2-methoxy-2-(4-methoxyphenyl)-ethylcinnamamide.

Toxicity

The leaves are said to cause abortion and sterility in women. The bark is used as a fish poison in the Celebes. Tannin, ingested frequently and in quantity over a long period of time, is antinutrient and carcinogenic.

Other Uses

Fruit: The fruit pulp has detergent action and has been used for washing clothes. Quisumbing says that bael fruit is employed to eliminate scum in vinegar-making. The gum enveloping the seeds is most abundant in wild fruits and especially when they are unripe. It is commonly used as a household glue and is employed as an adhesive by jewellers. Sometimes it is resorted to as a soap-substitute. It is mixed with lime plaster for waterproofing wells and is added to cement when building walls. Artists add it to their watercolours, and it may be applied as a protective coating on paintings.

The limonene-rich oil has been distilled from the rind for scenting hair oil. The shell of hard fruits has been fashioned into pill-and snuff boxes, sometimes decorated with gold and silver. The rind of the unripe fruit is employed in tanning and also yields a yellow dye for calico and silk fabrics.

Leaves: In the Hindu culture, the leaves are indispensable offerings to the 'Lord Shiva'. The leaves and twigs are lopped for fodder.

Flowers: A cologne is obtained by distillation from the flowers.

Wood: The wood is strongly aromatic when freshly cut. It is gray-white, hard, but not durable; has been used for carts and construction, though it is inclined to warp and crack during curing. It is best utilized for carving, small-scale turnery, tool and knife handles, pestles and combs, taking a fine polish.

Medicinal Uses: The fresh ripe pulp of the higher quality cultivars, and the "sherbet" made from it, are taken for their mild laxative, tonic and digestive effects. A decoction of the unripe fruit, with fennel and ginger, is prescribed in cases of hemorrhoids. It has been surmised that the psoralen in the pulp increases tolerance of sunlight and aids in the maintaining of normal skin colour. It is employed in the treatment of leucoderma. Marmelosin derived from the pulp is given as a laxative and diuretic. In large doses, it lowers the rate of respiration, depresses heart action and causes sleepiness.

For medicinal use, the young fruits, while still tender, are commonly sliced horizontally and sun-dried and sold in local markets. They are much exported to Malaya and Europe. Because of the astringency, especially of the wild fruits, the unripe bael is most prized as a means of halting diarrhea and dysentery, which are prevalent in India in the summer months. Bael fruit was resorted to by the Portuguese in the East Indies in the 1500's and by the British colonials in later times.

A bitter, light-yellow oil extracted from the seeds is given in 1.5 g doses as a purgative. It contains 15.6% palmitic acid, 8.3% stearic acid, 28.7% linoleic and 7.6% linolenic acid. The seed residue contains 70% protein.

The bitter, pungent leaf juice, mixed with honey, is given to allay catarrh and fever. With black pepper added, it is taken to relieve jaundice and constipation accompanied by oedema. The leaf decoction is said to alleviate asthma. A hot poultice of the leaves is considered an effective treatment for ophthahnia and various inflammations, also febrile delirium and acute bronchitis.

A decoction of the flowers is used as eye lotion and given as an antiemetic. The bark contains tannin and the cournarin, aegelinol; also the furocourmarin, marmesin; umbelliferone, a hydroxy coumarin; and the alkaloids, fagarine and skimmianine. The bark decoction is administered in cases of malaria. Decoctions of the root are taken to relieve palpitations of the heart, indigestion, and bowel inflammations; also to overcome vomiting. The fruit, roots and leaves have antibiotic activity. The root, leaves and bark are used in treating snakebite. Chemical studies have revealed the following properties in the roots: psoralen, xanthotoxin, *O*-methylscopoletin, scopoletin, tembamide, and skimmin; also decursinol, haplopine and aegelinol, in the root bark.

Cooperative's Role in Food Security

The World Food Summit Plan of Action stipulates that:

Governments, in cooperation with the private sector and non-governmental organizations, will:

- *Foster the social and economic organization of the rural population with particular emphasis on the development of small-scale farmers', fishers' and foresters' cooperatives, community organizations and development associations, so that rural inhabitants may be actively involved in decision-making, monitoring and evaluation of rural development programmes;*
- *Promote the empowerment of small-scale family farmers, fishers and foresters, both women and men, to set up their own cooperatives and business undertakings, as well as farmers' and fishers' financial and mutual institutions.*
(Commitment Three, Objective 3.5)

FAO and Cooperatives

FAO-ICA Collaboration

Addressing the NGO Forum for the World Food Summit in November 1996, the Director General of FAO stressed that today "everyone recognizes that governments alone cannot solve the problem of food security and if we are to make any progress we need the energy and expertise that reside in civil society."

In FAO´s view, as a generic term, "civil society organization" embraces a vast, heterogeneous and multifaceted "set of relational networks", and includes trade unions, self-help associations, cooperatives, women´s groups, development and advocacy NGOs, and informal groups alike. However, given this great heterogeneity, priorities have to be set and FAO thus pays special attention to membership-based, representative self-help organizations of farmers, fisherfolk and foresters, in particular their genuine cooperatives. These cooperatives, in spite of many failures and shortcomings, are traditional organizations of mainly the poorer segments of society which have the potential to play an important role in developing a strong "social capital" in rural areas that is regarded as a pre-requisite for food security and sustainable development.

Converting these concepts into activities and outputs, FAO focuses its cooperative assistance programmes on three technical areas:

- design of national policies, strategies and legislation to enhance cooperatives´ role in development, emphasizing the need for synergy and the economic efficiency gains to be derived from their involvement in decision-making and programme implementation;
- development and refinement of concepts and techniques for coalition-building and partnerships in support of rural development and food security;
- internal capacity-building to make cooperatives, along with other civil society organizations, equal and efficient partners in development.

After a meeting between Mr. Jacques Diouf, Director-General of FAO and the new President of the International Cooperative Alliance, Mr. Roberto Rodrigues, a Memorandum of Understanding has been finalized to ensure cooperation between FAO and ICA by consultation, exchange of information and coordination of efforts in fields of common interest, in accordance with the objectives and principles of FAO and ICA, with particular reference to the promotion of cooperative principles and methods in the areas of food, agriculture, fisheries and forestry, and related areas.

FAO provides continued technical and financial support, also through its Regional Office for Asia and the Pacific (RAP) in Bangkok, to the Regional Network for the Development of Agricultural Cooperatives in Asia and the Pacific (NEDAC). NEDAC was established with the assistance of FAO and has been supported by both FAO Headquarters and the Regional Office. Recent milestones of the FAO assistance include the following:

- NEDAC-FAO Regional Workshop on the Strengthening of Management and Development of Agricultural Cooperatives, organized in collaboration with the Cooperative Development Authority (CDA) of the Philippines and held in Manila from 19-25 April 1997
- the Third Session of NEDAC's General Assembly was held with FAO technical and financial assistance in New Delhi, India from 24-28 November 1997
- FAO provided technical and financial support for the organization of the NEDAC Workshop on Financial Viability of Agriculture Cooperatives, held in Comilla, Bangladesh, from 20-23 May 1998
- FAO sponsored the "FAO-NEDAC Training of Trainers Workshop on Agricultural Cooperative Development" held at the University of the Philippines Los Banos, College of Public Affairs, Laguna, Philippines, from 26 to 30 October 1998.

Agricultural Cooperative Development

Globalization, trade liberalization, deregulation and changes in political and economic paradigms have led to major changes in the political, social and economic environments within which agricultural cooperatives operate. The decline of government assistance and growing competition from the private sector are now forcing cooperatives to reassess their performance and seek ways and means to improve their business efficiency in providing better member services

In the framework of its cooperative training of trainers programme aimed at helping developing countries and countries in transition transform their agricultural cooperatives into genuine self-help organizations, FAO has published in English an innovative manual geared to encouraging greater membership participation, improving management and familiarizing political and administrative decision-makers with new cooperative development approaches. The training manual is composed of seven modules, as follows:

- Understanding Cooperatives,
- Participation & Learning,
- Communication,
- Organizational Development,
- Cooperative Management,
- Tools for Planning and Organizing Cooperative Activities,
- Participatory Appraisal, Monitoring and Evaluation.

Throughout these modules, the manual deals with ways in which trainers and promoters of cooperatives can support cooperative members and management in the development of their cooperative organizations. Its objectives are:

- to broaden the trainers´ view of their role, and ways in which they can react to circumstances, and to increase their confidence in their own capabilities. In short, to increase their competence.
- to help the trainer become an effective facilitator and moderator. That is to say, someone who can offer new methods for dealing with problems and tasks, assist in solving conflicts, draw attention to alternatives and assist in the more effective operation of the cooperative.
- to equip the trainer with tools to act as a resource person, providing information to the cooperative organization, its members, leaders and managers.
- to familiarize the trainer with participatory techniques which involve all parties concerned with the future of cooperative organizations.
- to support the trainer in the various tasks of raising awareness among cooperative promoters, members, decision-makers, leaders and managers of their problems/ constraints as well as in their potential/capabilities, without taking the initiative away from them.

The manual will guide trainers through the process of training by providing a standard structure in each module, including the following units: Objectives of the unit, Key learning points, Teaching strategy, and Reference information. A range of examples and exercises are also given. The English version of the training manual, which was field tested in selected countries in Asia and Africa before printing, is now being disseminated. At same time, preparation of the French version through a process of translation, adaptation and testing has started and will be completed later this year.

Completion of FAO Study

This study on "Capital Formation and Kenyan Farmer-Owned Cooperatives" was carried out in collaboration with the Turku School of Economics and Business Administration, Turku, Finland and thanks to extra-budgetary financing received from the Government of Finland.

Field research was conducted in Kenya during 1998 and focused on an in-depth examination of capital formation and investment behaviour in a sample of eight cooperatives within Kenya's dairy and coffee sectors. The detailed analytical report of 325 pages, including appendices, contains valuable insights on some of the main constraints in strengthening the capital formation and business performance of agricultural cooperatives in Kenya under current liberalized market conditions, and provides a series of useful recommendations to address these concerns in the sub-region.

FAO is now in the process of publishing a summarized version of the study findings for broader dissemination within the sub-region.

FAO has assisted in:

- Capacity-building for agricultural cooperative development in Yemen (1999, completed)
- Developing agricultural producers' unions in Turkey (1999, ongoing)
- Comparative policy and programme development in Vietnam (1998, ongoing)
- Preparing for the restructuring and institutional strengthening of the cooperative sector in Morocco (1998, ongoing)

- Analyzing the potential role of cooperatives and rural finance in an EU accession environment (1998, ongoing)
- Organizing a national workshop on training of trainers in cooperative membership development in Slovenia (1997)
- Revitalizing the Regional Network for the Development of Agricultural Cooperatives in Asia and the Pacific (NEDAC) (1997)
- Training of trainers in cooperative development in Ethiopia (1996, 1997)
- Enhancing farmers' participation in cooperative development in Tanzania (1996)
- Conceptualizing and applying the "Systemes de-gestion appropries des cooperatives de petits exploitants agricoles (GACOPEA)" training approach in Burkina Faso (1996)
- Strengthening the capacity of the Office de Development de la Cooperation (ODCO) in Morocco (1996)
- Agricultural cooperative building and training in the North of Thailand (1995-96)
- Strengthening cooperatives' role in horticultural development in Madhya Pradesh, India (1995-96)
- Developing a new cooperative system in agriculture in Vietnam (1995)
- Organizing an international workshop on cooperative legislation in China (1994)
- Promoting cooperative "reconversion" programmes in the MERCOSUR countries in Latin America (1994, ongoing)
- Strengthening the capital base of rural cooperatives in India, Tanzania, Kenya and Guatemala (1993, ongoing)
- Testing small group-and cooperative-based participatory training approaches in collaboration with the Cooperative College, Moshi, Tanzania (1992-95)
- Developing new legal policies encouraging healthy independent rural cooperative growth in Ethiopia, Guinea, Tanzania, Vietnam and India (1990-96)
- Building sub-village self-help groups in the framework of its People's Participation Programme in Ghana, Kenya,

Lesotho, Nicaragua, Pakistan, Sierra Leone, Sri Lanka, Tanzania, Thailand, Zambia and Zimbabwe (1982, 1997)

- Organizing series of workshops on the problems and strategies of the transformation of agricultural cooperatives in the transitional economies of Central and Eastern Europe (1992-95)

Formation and Management Training

This workshop was organized in collaboration with the ICA Sub Regional Office for Eastern, Central and Southern Africa and held on January 18-23, 1999 at the Uhuru Lutheran Hostel in Moshi, Tanzania. The main purpose of this workshop was to discuss the findings and recommendations of the FAO-Turku study on agricultural cooperative capital formation in Kenya and to critically review the draft version of the FAO Training of Trainers manual on agricultural cooperative development for its adaptation within the sub-region.

Twenty-five persons from Ethiopia, Kenya, Tanzania, Uganda, and Zambia participated in the workshop, including government officials, cooperative trainers and movement leaders. They came from government ministries, cooperative institutions, and national cooperative federations, respectively.

The others were representatives from the co-sponsoring agency, *i.e.* the organising International Cooperative Alliance (ICA), from FAO Headquarters, from the FAO Regional Office for Africa and international researchers from the Turku School of Economics and Business Administration.

The workshop was divided into two working sessions: the first two-and-a-half day session addressed the theme of cooperative capital formation in agricultural cooperatives in the sub-region and lasted from Monday, 18 January until mid-day, Wednesday, 20 January. The second session on the topic of cooperative management training began on Thursday, 21 January and lasted until mid-day Saturday, 23 January.

Workshop *recommendations on capital formation* included:

- the need for a critical review of cooperative principles, legislation and policies in the light of recent market liberalization within the sub-region

- the need for a critical review and reform of existing taxation policies affecting agricultural cooperatives in the sub-region
- the need for improved training in the field of cooperative financial management
- the need for modernization and computerization of agricultural cooperative business information systems

Workshop *recommendations on cooperative training* included a series of technical suggestions to improve the content of the manual and concrete suggestions on how to better adapt the manual to conditions within the sub-region.

Fertilizer Cooperative

The Indian Farmers´ Fertilizer Cooperative Limited (IFFCO), the largest fertilizer sector cooperative, was established in 1967, with the primary objective of increasing agricultural productivity and improving the rural economy by producing quality fertilizers and promoting the balanced use of these fertilizers. It is a multi-unit cooperative-a federation of approximately 30,000 societies from village to national level. IFFCO has also established 174 Farmers Service Centres where fertilizers, seed and agro-chemicals are supplied under one roof and where farmers receive technical advice on the use of these agricultural inputs. The cooperative has far-reaching national linkages with farmers, researchers, extensionists, the private sector and government.

Since 1992, IFFCO has launched an Integrated Plant Nutrition System (IPNS) programme in response to the identified need for more balanced nutrient application in Indian agriculture. The main objective of the programme is to improve soil fertility and productivity at farmers´ level in order to increase agricultural production and farmers´ income.

FAO has been involved with the IFFCO IPNS programme since its inception in 1992, regularly providing advice on various technical aspects of its activities. A joint workshop on IPNS was organized in September 1992 and in 1993 the cooperative launched a programme of trials and demonstrations to promote IPNS. FAO.

Consultants have visited selected villages and established plant nutrient balance sheets for those villages. As a result of those balance sheets, FAO has been able to propose improved plant nutrient management techniques to farmers in those villages, thereby improving their productivity and fertilizer use efficiency and reducing nutrient losses to the environment.

In April 1996, FAO agreed with the Government of India to finance and provide technical backstopping to a Technical Cooperation Programme (TCP) entitled "Development of an Integrated Plant Nutrition System Methodology". IFFCO has been actively involved in implementing this Programme. In September 1996 IFFCO organized a workshop for all its regional staff involved in balanced fertilization and the IPNS programme throughout the country, as well as university professors and experts from the Indian Council for Agricultural Research (ICAR) and the Fertilizer Association of India (FAI). In this workshop the results of the last six IPNS campaigns were discussed and plans for future activities were elaborated. The most significant activity under the IFFCO-FAO Programme was the organization of an International IPNS seminar in November 1997, in which researchers involved in IPNS work from both the developed and developing world met to share their experiences in this area of work. Another important output of this TCP is a publication entitled "A Guide to Field Implementation of IPNS". The technical assistance continues with the framework of a Letter of Agreement arrangement and catalytic FAO funding support.

The Role of Cooperatives

FAO´s Fishery Resources Division and Rural Development Division has launched a joint project which aims to prepare a "source book" on the role of small-scale aquaculture within rural development, with special reference to existing forms and models of cooperatives and other farmers´ organizations in aquaculture, and the potential role and forms of cooperatives/producers´ (farmers´) associations in managing aquaculture resources. During the first phase of project implementation in 1998 and early 1999, five case studies were prepared on the subject matter in Kenya, Laos, Panama, Sri Lanka and Zambia.

Currently, a *comparative analysis* of case the studies and *consolidated assessment* of the role of cooperatives and other forms of associations in managing aquaculture resources is being carried out, the findings of which will be published later this year.

Continued Collaboration with COPAC

FAO, one of the founding members of the Committee for the Promotion and Advancement of Cooperatives (COPAC), which also hosted the COPAC Secretariat between 1971 and 1996, continues to provide its technical and financial support for the operation of this unique international cooperative organization. The World Food Summit Plan of Action and FAO´s new policy puts a strong emphasis on the importance of building new partnerships, networks, and coalitions among governmental and non-governmental organizations for sustainable agricultural and rural development and food security. COPAC is considered one of the best examples of successful networks at international level, which was established far before the issue became such a high priority on the global agenda.

In light of FAO´s new policy of strictly linking expenditures to outputs, increasing importance is attached to the complementarities in the activities of FAO and COPAC to make sure that COPAC will produce important outputs which, without COPAC, FAO would not be able to produce. In 1999, the main areas of collaboration between the Rural Development Division of FAO and COPAC are as follows:

Preparation of a typology of and lessons learnt from existing strategic alliances of cooperatives, farmers´ and rural workers´ organizations at grassroots level and proposals for enhanced strategic alliances.

As a follow-up to the World Food Summit, FAO´s Rural Development Division has been actively involved in the promotion of the ACC Network on Food Security and Rural Development at both international and national level. At national level, Thematic Groups on Rural Development and Food Security have been established in the framework of the Network, in which international and national public and civil society organizations work together to enhance synergies in their activities for sustainable food security. The scheduled COPAC study will contribute to extending the scope of Thematic Groups´ to grassroots level by providing a series of case studies on successful examples of strategic alliances among the primary organizations belonging to the constituencies of COPAC, such as agricultural and other rural cooperatives, credit unions, farmers´ and workers´ organizations, trade unions.

COPAC Open Forum on Successful Cooperative Development

In 1998, FAO´s Rural Development Division and its Sub-Regional Office for Central and Eastern Europe launched a joint project entitled "Institutional Patterns of Rural Development in the Countries of Central and Eastern Europe". The long term objectives of the ongoing project is to develop post-transitional institutional models of rural development for the specific conditions of the former centrally planned economies of Central and Eastern Europe. Cooperatives call for particular attention in the process of institutional transformation in the sub-region. The theme of the COPAC Open Forum is fully relevant to the FAO project mentioned above. COPAC will arrange for a number of unbiased country case studies on successful surviving and new cooperative forms for discussion at the Forum, the findings of which will be consolidated in a synthesis report.

Cooperative Legislation and Public Policy

FAO has a significant programme providing legal assistance to member states on cooperatives mainly in the framework of the Rural Development Division´s regular programme, aiming to prepare and disseminate guidelines for member governments on restructuring the cooperative sector in the context of market-driven reforms. In addition, assistance by the Development Law Service centres on the drafting of agricultural legislation, but legal advice is also given in connection with institutional restructuring, international agreements and other matters. COPAC will organize a consultation entitled *Cooperative Legislation and Public Policy,* which will include discussion of and a report on timely legal aspects of cooperative restructuring processes in developing countries and countries in transition.

COPAC Web Site and Databank Project

In the framework of these COPAC activities, FAO supports the establishment of an interactive databank of useful agricultural cooperative business/trade information, including agricultural market analysis and trade forecasts which could be accessed by agricultural cooperatives in developing countries, in particular in the Africa region. This interactive databank complements FAO´s current efforts aimed at strengthening agricultural cooperative capital formation and investment decision-making in that region and is a logical follow-up to recommendations arising out of the recent sub-regional workshop on that topic. Furthermore, the COPAC web pages include links to FAO information on cooperatives-highlighting new publications on the Rural Administration and Cooperatives, and agricultural cooperative legislation on FAOLEX database, and identification of cooperative development projects undertaken by members and non-members to allow COPAC members to pursue joint cooperative development activities.

The Gender Dimension

The world over, statistics show that women´s participation in cooperatives is low, especially in rural cooperatives. Cooperatives have tended to be synonymous with "men´s" cooperatives. Rural women's participation in agricultural cooperatives and their decision-making processes needs be increased, particularly in view of the important activities women carry out in the strive for food security through agricultural production, and the impact cooperatives have through its importance in their village life.

Cooperative law often condones such discrimination by providing that the head of the family attends meetings: the fact that the wife is often *de facto*-or even *de jure*-head of the family is not always seen as enough reason for her to participate. When cooperative laws are revised, all provisions which make for gender discrimination should therefore be weeded out to avoid aggravating the problems faced by women in their attempts to be integrated into the participatory cooperative structures.

In addition, with greater freedom to decide on the types of business to conduct through a cooperative, the way is open for the development of activities of specific interest to women such as small cooperative mills, food storage and preservation, production of household necessities like soap and clothing, small animal raising and handicrafts. More stress should also be laid on cooperatives´ social function by organizing services which would relieve women from certain of their tasks: child care services or drudgery-reducing activities, or assist with organizing marriages and other ceremonies.

Following the request from the Government of Bulgaria, FAO provided assistance under its Technical Cooperation Programme (TCP) to the formulation of a strategy for agricultural development and food security consistent with sustainable economic development, GATT/WTO commitments and the country's objective to join the EU. Given the long history of rural savings and credit cooperatives in the Bulgarian economy, an analysis of the present and future role of the Private Mutual Rural Credit Associations (PMRCAs) as a provider of durable and sustainable financial services to the agricultural and rural sector was carried out.

Strategy to Combat Drought and Famine

During the previous century, the Indian arid zone experienced agricultural drought in one part or the other during 33 to 46 years which suggests a drought once in three years to alternate year. Often drought persists continuously for 3 to 6 years as prolonged drought faced by this region during 190305, 1957-60 1966-71,1984-87 and 1997-2000. Such prolonged droughts put tremendous stress on natural resources arid lead to severe scarcity of food, fodder and water.

The major causes of agricultural droughts in the Indian and zone are its geographic location not favouring abundant monsoon rainfall, poor quality and excessive depth of ground-water limiting its use for irrigation, absence of perennial rivers and forests, poor water holding capacity of soils, arid huge drawal of limited ground-water resources. Because of lack of Substantial irrigation facilities, the agriculture is mostly dependent on rainfall. The increased pressure of both human (400%) and livestock (127%) population during twentieth century has put tremendous pressure on land, and Surface and ground-water resources. Therefore, the impact of drought is felt much more severely in the arid region compared to other parts of the country.

As the water storage is dependent on the scanty and erratic rainfall, the duration of availability of water in Surface water resources is reduced significantly in drought years. Ground-water table is declining @0.2 to 0.4 m/annum in almost three-fourth of the region, consequently shallow wells dry up during droughts and deep wells become deeper. Quality of ground-water deteriorates and sometimes the concentration of undesirable substances such as fluoride and nitrate increase to harmful/toxic levels. Grazing herds of animals quickly remove the scanty grass cover that come up with meagre rainfall, thus aggravating the problems of soil erosion and desertification. Widespread crop failures lead to acute shortage of food and fodder. Both human and livestock suffer from malnutrition and consequently become victim of host of diseases. As most of the people of this region depend on agriculture and pastoralism, drought leads to decline in income and employment opportunities. Large-scale migration with livestock or in search of employment is a common feature during prolonged droughts.

To minimize the Suffering of human and livestock, relief measures are taken Lip by the respective State Governments and NGOS' on a large scale. These measures mainly aim at provision of drinking water, supply of food grains through Public Distribution-System at subsidized rates, supply of food arid fodder for livestock arid Subsidy to approved Gosalas, human and livestock healthcare, etc. Efforts to create direct and indirect wage employment through food for work barely sustain the living of the rural poor who suffer most due to drought. However, some long-term preventive measures need to be given increased attention for integrated development of drought-prone areas and to tackle the problem on permanent basis. Some of these direct measures are long-term forecast of monsoon, suitable lad use system, water harvesting, soil and water conservation, contingency crop planning, adoption arid improved technologies for dryland crops, efficient irrigation methods, enrichment of cereal straw as fodder, etc. In addition to these, human and livestock Population pressure needs to be reduced through education, alteration ways of employment generation arid disposal of unproductive livestock.

The root cause of weak monsoon in India is related to the widespread, persistent atmospheric Subsidence, which results from the general circulation of atmosphere. Better understanding and mathematical modelling of the monsoon phenomenon would be very helpful in early long-term forecast of monsoon to enable planners arid farmers to plan accordingly.

No one understands the value of single drop of water better than the desert dwellers. Rainwater harvesting is traditional way of life in arid regions. Various techniques of rainwater harvesting have been developed/refined by research workers. Improved designs of Nadis, Tankas, Khadins, etc. have also been developed. These technologies should be popularized among the people of this region. Utilizing flash floods/surplus rainwater for artificial recharge of ground-water to augment the dwindling water table is need of the hour. Integrated watershed management, which aims at utilizing the rainfall wherever it falls should be the unit for planning and implementation of the development programmes. The measures like afforestation, pasture development, livestock management, field crops, water Storage, etc. are undertaken in the watershed areas identified as suitable for Such measures.

With the increased pressure oil land, marginal lands are being brought under cultivation, which is a disastrous trend. Concerted efforts have to be made to adopt Suitable land use systems keeping in consideration tile rainfall, soil type and need of the people. Growing of crops, fruits, trees and grasses various combinations minimize the risk of crop failure and provide stability to farm income. Suitable combinations of these components for different rainfall zones and soil types have been developed by CAZRI Scientists which should be preferred over crop Cultivation alone. As water is the scarce resource in the Indian arid zone, efficient irrigation technologies like sprinkler and drip system should be Popularized which aim at minimizing production per unit of irrigation water. Adoption of improved agronomic practices like use of improved varieties, timely weed control, use of fertilizers along with farm yard manure, in-situ rainwater harvesting, etc. can give good yields even in below normal rainfall years. Cultivation of water intensive crops should be discouraged.

Besides natural resources, livestock and permanent Vegetation Such as grasses and trees are strengths for Survival of mankind in the arid regions. Management of grasslands with Lasiurus Sindicus, Cenchrus ciliaris arid Cenchrus setigerus and top feed specie Such as Prosopis cineraria, Acacia sengal and Tecornella undulata need priority attention. Such a silvi-pasture system survives annual droughts and provides rich fodder.

Quality of fodder particularly the wheat straw given to cattle during drought is usually very poor. The fodder quality can be improved through urea/molasses treatment, thus improving animal health and productivity with very little investment. Management of common property resources such as grazing lands, dams, village ponds, etc. needs top priority by the people for themselves to combat drought in and regions.

The drought situation like the present one during the past three year (1999, 2000, 2001) to occur in this region and it needs a proper planning to combat drought/famine. Drought impacts are generally more severe on livestock than on human beings. Some of the measures that can be taken up for combating drought could be as follows:

Short-term measures:

1. Early warning and drought monitoring should be carried out on the basis of long, medium and short-term forecasts.
2. Constitution of task forces in each district to initiate relief measures immediately after the drought strikes.
3. Supply of good quality drinking water to human and livestock in severely affected areas.

4. Fodder banks should be established in the region. Low quality fodder/alternate fodder resources should be enriched to meet the protein demand.
5. Cattle camps should be opened and fodder should be provided at a subsidized rate.
6. Contingency crop plans should be prepared in advance to meet out the aberrant weather conditions such as early/late setting of monsoon and or early/late withdrawal of monsoon.
7. Implementation of crop and livestock insurance schemes.
8. Timely availability of credit, postponement of revenue collection and repayment of short-term agricultural loans.
9. Training of personal involved in drought relief measures on short/long-term basis.

Long-term measures:

1. Greater coordination among line departments should be ensured.
2. Dependence on animal husbandry and silvipasture system for alternative source of income in arid region.
3. Rainwater harvesting for both the drinking and in-situ cropping, improvement and popularization of traditional rainwater harvesting systems arid rainwater conservation/efficient utilization.
4. Rejuvenation of traditional rainwater harvesting systems *viz.* Nadis, Tankas, Khadins, etc.
5. Systematic study on the use, artificial recharge and augmentation of ground-water aquifers.
6. Completion of IGNP lift canals to divert water to drought prone areas.
7. Human and livestock population should be managed to reduce the pressure on fragile arid ecosystem.
8. Popularizing the improved agronomic practices to maximize the crop yield per raindrop.
9. Integrated watershed management for efficient management of land and water resources should be given top priority. Appropriate land use planning discouraging

water intensive crops, encouraging sprinkler and drip irrigation systems, and practising alternate land use such as agro-forestry, agro-horticulture and silvi-pasture would provide long-term drought proofing in the Indian arid zone.

Benefits of Global Warming

I just returned from an extended visit to India, my country of birth, and was astounded by the rapid economic progress India has made in the last 25 years. My recent trip was after a gap of five years, a 'long' gap considering that in last five years India has maintained a steady growth rate of about 7% and India today is poised to become the third largest economy in the world (after USA & China) on a PPP (Purchasing Power Parity) economic model by the end of 2006. India today has a population of about 1050 million (more people live in India than the total population of Europe including Eurasia), is a strong and free democracy and is enjoying an unprecedented prosperity, thanks to the growth of knowledge-based economy like IT (Information Technology) and Bio-medical technology plus a strong consumer-oriented economy driven by about 300 million plus young Indians with increased disposable income to spend on food, clothing and various other consumer goods. As a weather & climate scientist, what impressed me was the fact that India's strong economic progress has come about in an increasingly warmer world of the last forty years or so, completely defying the projections of deleterious impact of Global Warming by IPCC (Intergovernmental Panel for Climate Change, a United Nations Group of Scientists) and its supporters.

My first regular job in India was in the India Meteorological Department in Poona (Now Pune) where I worked as a statistician doing climate summaries and related work in the late fifties. I was briefly involved in doing statistical analysis of a heat wave of 1956-57 which killed several hundred people in Northern India and soon became interested in the weather & climate of India and in seasonal prediction of Indian Monsoon. During my student days in Pune in early to mid-fifties, I remember distinctly that the city of Pune (with a population of about half a million then, always enjoyed moderate climate because of its location in the foothills of the Western Ghats at an altitude of about 600 m) rarely recorded a maximum temperature over 35C.

Today, Pune and vicinity has a population of close to five million and with huge development of concrete buildings and other infrastructure, the city records maximum temperature of 37C and above for at least ten days or more, from Mid-March to Mid-May. Other cities in central India like Hyderabad (population about 6 million today) routinely record maximum temperature of 40C and over for several afternoons during the pre-Monsoon months of March to May. In New Delhi, India's Capital city (latitude ~30N), with a population of over 10 million people, maximum temperature of 45C during the pre-Monsoon months is more common today than fifty years ago. Most of this increase in the mean maximum temperature at most mid-size to large cities in India has come about, primarily as a result of increased urbanization and land-use change. According to a study by Roop Kumar et al (1994), mean temperature over India as a whole has increased by about 1C in last fifty years or so. However, this increase in mean temperature has not adversely impacted India's economy or its people. India has significantly improved its grain, vegetable and fruit production in last fifty years. Let us look at some of the numbers:

India's rice production has grown four-fold, from 25 M tons in 1950 to about 100 M tons in 2000 (Selvaraju 2003) and most of India's rice production is in the southern region of India where mean temperature has increased by 1C. In Northwest India, winter wheat production has grown almost five-fold, from 12 M tons (1950) to about 60 M tons (2000), thanks to the Green Revolution inspired by the Nobel Peace laureate Norman Borlaug and improved irrigation technology in the state of Punjab. Production of fruits like bananas and mangoes has grown significantly in last 25 years, thanks to improved technology leading to growth of high-breed fruit varieties, suitable for a warmer climate. Indian fruits are enjoying increasing export to countries in the Middle East and Europe, while within India fruit packaging and processing industry has grown substantially. Supply of fresh vegetables and spices has also increased substantially in recent years (India is primarily a vegetarian country).

India today is the largest milk producing country with over 80 million tons of milk a year ensuring a livelihood for about 15 million farmers and their families in close to 100,000 village-level societies across the country. The bulk of India's milk is provided by water buffaloes, a docile ubiquitous herbivorous animal, slightly bigger than an average cow but fetching more milk than an average cow, with 6% or higher butter-fat content. Recent advances have helped develop high-breed buffaloes which can yield up to 20 liters of milk per day and these buffaloes are well suited to thrive in a warmer climate. Increased milk production has helped India's dairy industry to produce nutritious food for the younger generation (there are an estimated 200 million young Indians under the age of 15 today).

The Indian Monsoon is the most important climate event in India and its interannual variability and linkage to large-scale atmosphere/ocean circulation patterns has been the subject of hundreds of studies in last fifty years. The summer Monsoon (June-September, the wet season) provides the bulk of rainfall for most of India, while the winter Monsoon (December-February, the dry season) provides rainfall only to the southeast regions of the Peninsular India. Climate modelers and IPCC documents have projected a more vigorous global Monsoon and a more variable Asian Monsoon circulation in a warmer world. In reality, observed global monsoon circulation has weakened in the last fifty years (Chase et al 2003) while Indian/Asian Monsoon circulation, while showing decadal variability-from below normal to above normal over a 30-year period-shows no discernible impact of global warming at present (Kripalani et al 2003). Decreasing summer Monsoon rains of last several years have aggravated water shortage situation in many regions, especially in the western State of Gujarat, where an ambitious project to divert waters of the Narmada River (which flows into the Arabian Sea near the Gulf of Khambhat) to the arid regions of north Gujarat is underway at present. Preliminary success of this project has prompted other local and regional governments to take up similar projects on other rivers, especially in southern India where most rives become 'bone-dry' following the summer Monsoon season. Providing adequate water for a billion plus populace of India is perhaps the most challenging problem for Indian scientists and policy-makers at present. Numerous research and development projects are underway at present in many Universities and Technological Institutes on Monsoon variability, water resources and related issues.

On the social and cultural scene, India's Bollywood in Mumbai has become the world's biggest producer of full length movies and the movie industry while employing large number of young artists and technologists is making increasing use of weather & climate themes by depicting winter scenery of Himalayan foothills along with waterways of Kerala State and beautiful beaches of Goa. The ever growing popularity of Indian movies has helped India's tourism industry which has grown significantly in recent years with young Indians exploring the breadth and length of India and its climatic zones from the mid-latitude like climate in the Himalayan foothills to the Equatorial Trough like climate in the southernmost states of Kerala and Tamil Nadu.

In summary, India today has done well by adapting to the warmer climate and associated climate change. Indians do not mind hot weather, in fact they thrive on it! The India Meteorological Department has increased its mandate and personnel substantially in recent years and has developed improved capabilities for foreshadowing of the summer Monsoon rainfall and associated variability. Extreme weather events like Bay of Bengal cyclones striking coastal regions of southern India are monitored at present with improved technology and this has resulted in reduced human and property losses in recent years.

Workshop on "Nutrient Status, Need And Recommendations for Major Fruit Crops of Uttar Pradesh and Uttaranchal"

India is endowed with varied agro-climatic conditions and enjoys an enviable position for growing almost all types of horticultural crops (tropical, subtropical and temperate). At present, it is one of the largest producer of fruits. Its production has increased 5 to 8 times since independence, *i.e.*, from 5.5 million tonnes (MT) in 1952 – 53 to 45.5 MT in 1999 – 2000. The productivity of fruits per unit area in India has also increased from 10 to 12 MT per ha in almost one decade. India contributes 10.2 per cent to the total world production of fruits.

Globalization has thrown up a lot of opportunities and challenges for enhancing Indian trade of fruits. India, being a leading producer of fruits, can establish its position on a sound footing as a major player in the international market. Among several factors, one of the driving forces behind increased productivity is the use of mineral fertilizers. However, the average productivity of most of the fruit crops is far below their potential. Lack of genuine planting material, lack of adequate infrastructure, poor crop management system, etc., are some of the impediments. In India, Uttar Pradesh and Uttaranchal are the only twin states where a large number of subtropical and temperate fruits are grown, but due to lack of proper agro-management, appropriate fertilizer schedule and realization of full genetic potential, productivity remains low. Therefore, to increase production, an urgent need to review and refine nutrient recommendations for major fruit crops of these two states has been felt. With the advent of WTO regimes and India being one of the signatories, it has become necessary to device strategies to produce quality products at a very competitive cost.

The decision of holding a workshop on review, refinement and recommendations of nutrients for major fruit crops under the aegis of Central Institute for Subtropical Horticulture (ICAR), Lucknow in collaboration with Potash Phosphate Institute of Canada – India Programme, Gurgaon during Dec. 10 – 11, 2002 is indeed very timely. I am delighted to learn that the two day event was a grand success and provided the much needed platform for interaction between experts of nutritionist of scientific organisation and progressive farmers as the relevant topical issues were presented by eminent experts. Proceedings of the workshop on nutrient management will exemplify the role of fruit crop nutritionists efforts for the development of long term use of fertilizers in these crops.

The main purpose of the book is to review the present status of nutrient requirement and fertilizer application and being into focus the recommendations made thereof for higher production of fruit crops. It is hoped that this publication will be of immense help to the farmers, scientists, planners, research institutions, agricultural universities, academia as well as the general reader eager to know about nutritional requirement of fruit crops. The present book consists of detailed chapters dealing basically with review and refinement of nutrient recommendations along with their physiological effects on major fruits crops like mango, guava, aonla, papaya, litchi, citrus, pears, peaches, etc., grown in Uttar Pradesh and Uttaranchal.

Terminator of Food Security

'Terminator' technology, which could deal a fatal blow to millions of resource-poor farmers in developing countries, has attracted severe criticism across the world.

A Biotechnological innovation patented in the United States portends great danger to agriculture and food security worldwide, particularly in developing countries. Scientists have questioned the technology on ethical as well as scientific grounds. The patent, U.S. Patent Number 5,723,765 and titled "Control of Plant Gene Expression", was granted on March 3, 1998 by the United States Patents Office to the U.S. Department of Agriculture and the seed company, Delta and Pine Land Co., a subsidiary of the seeds and agro-chemicals multinational Monsanto/American Home Products. It is for a technique that can alter seeds genetically so that crops raised using them do not yield seeds that germinate.

The technology has been appropriately named "Terminator" by the Canadian non-governmental organisation (NGO) Rural Advancement Foundation International (RAFI), which has spearheaded an international campaign against it. Terminator alters the expression of certain genes in plants so that plants terminate their reproductive switch, abort the embryo and make themselves sterile. Such plants then produce seeds that cannot germinate. Although it may take a few years for this technology to become commercial (according to Monsanto's

Indian division, the gene is only a concept), for 90 per cent of the 100 million farmers of India, who traditionally save a portion of their harvest as seeds for the next sowing season, import of seeds with this gene would mean buying seeds every crop season from the seed companies. According to scientists across the world, Terminator could deal a fatal blow to some 1.4 billion resource-poor farmers in developing countries.

The utter disregard of the patent holders for farmers in developing countries is obvious from the justification offered by Murray Robinson, president of Delta and Pine Land Co., that the technology was devised to prevent theft and unauthorised cultivation of proprietary seeds. "It will provide seed companies a safe avenue for introducing this technology into countries such as India, Pakistan and China," Robinson said.

By Special Arrangement

Transgenic systems will have an adverse impact on traditional farming practices.

It took four years and an investment of $720,000 to develop the technique, according to RAFI. The USDA spent $190,000 and Delta and Pine Land Co. $275,000 on in-house research expenses and the two jointly spent $255,000. The patent covers seeds of all crop varieties, including genetically modified and conventionally developed seeds. Although the technology has been experimented only on cotton and tobacco, Murray Robinson believes that it will be effective in all major crops, including wheat and rice. While Delta and Pine Land Co. has been given exclusive licence to sell the seeds, the USDA is to get a royalty of 5 per cent on the net sales. The USDA and Delta Pine Land Co. have applied for licence in 78 countries. If the technology becomes widely licensed, the commercial seed sector would end up controlling the international seed market, including the market for the seeds of self-pollinating crops such as wheat, rice, soyabean and sorghum, which it has found difficult to control until now.

Delta and Pine Land Co. is the largest cottonseed company in the world, and had a turnover of $183 million in 1997.

Monsanto, its parent company, operates in India in the seeds and agro-chemicals market. Monsanto, which bought Delta and Pine Land for $1.76 billion a few weeks after the latter got the patent for Terminator technology, is now the largest pesticide firm, the second largest seed company and one of the 10 largest producers of pharmaceutical and veterinary medicines in the world.

Agricultural scientist Dr. M. S. Swaminathan, who was instrumental in making the Green Revolution a success in India, describes the technology as "unethical" and fears that it has the potential to "endanger the country's food security". He points out five major issues relating to the technology, which have caused concern.

First, he questions the ethical and social relevance of the technology in a country like India where the farming community comprises 70 per cent of the population and 90 per cent of the farmers depend on saved seeds and rely on the age-old practice of choosing seed varieties to suit the soil conditions. (In the U. S., hardly 2 per cent of the population is engaged in agriculture.) The Terminator mechanism would leave traditional farmers unable to use or exchange saved seeds.

Secondly, there is the danger of this technology affecting unintended targets. It is possible that pollen from crops grown using Terminator seeds would transfer the mechanism to crops grown nearby and cause sterility in their seeds too, which would be known only when they are sown the next season. This could lead to a fall in agricultural production in successive years, endangering the food security of the country.

The technology can affect the livelihood of the majority of Indians. Although only 30 per cent of India's Gross National Product (GNP) comes from agriculture, 70 per cent of its people depend on agriculture for their livelihood. So, in human terms, if not in economic terms, agriculture is very important for the country. According to Dr. Swaminathan, "the country's agriculture, livelihood and food security are at great risk from this killer technology."

The third concern is that it is possible to kill the embryo, in order to render the seed sterile, only by releasing some chemicals or toxins into it. As the seed is used only for consumption, what are its nutritional implications? According to Dr. Swaminathan, many diseases such as cancer occur owing to chance consumption of certain chemicals over a period of time. Thus, one is not sure if it is safe to consume grain that has been genetically engineered in this manner.

Even in the case of cotton, this technology can prove harmful. In India, cottonseed cake is used as animal feed, which means the Terminator chemical can get into the plant-animal food chain.

The fourth fear is that the country's rich genetic bio-diversity will be lost. Large areas may be covered by just one genotype, and would affect the efforts directed towards the revitalisation of on-farm conservation traditions and the breeding of location-specific varieties through participatory breeding methods. If this technology gains wide acceptance, farmers would have to buy seeds every year from the same company. There will thus be just one kind of seed everywhere for a given crop. Such homogeneity would enhance the seed's genetic vulnerability to pests and diseases.

The fifth concern is that if seed companies use tetracycline to activate the toxin gene, every seed sown will have to be treated with the chemical. What will be the impact of planting tetracycline-soaked seeds on soil ecology, particularly on micro-flora and fauna?

Scientists the world over have protested against the patenting of this technology, describing it as a major assault on resource-poor farmers. Britain's Prince Charles, who is against genetically-modified organisms (GMOs) and who has been practising organic farming for the last 12 years, cautions that "once such genetic material is released into the environment it cannot be recalled." In the July issue of the British magazine *The Plant Talk*, he says: "If something goes wrong (with such GMOs), we will be faced with the problem of clearing up a kind of pollution that is self-perpetuating. I am not convinced that anyone has an idea of how this could be done or who would pay for it...I personally have no wish to eat anything produced by genetic modification, nor will I knowingly offer this sort of a produce for my family or guests." Farmers' organisations across India have demanded a ban on any seed material containing the Terminator gene. The Karnataka Rajya Raitha Sangha even sought the termination of the research projects undertaken by the Indian Institute of Science, Bangalore, with Monsanto.

When news about the killer technology broke on the Internet, Dr. Swaminathan alerted the Director-General of the Indian Council for Agricultural Research (ICAR), Dr. R.S. Paroda, and the Secretary of the Department of Biotechnology (DBT), Dr. Manju Sharma, besides warning the Consultative Group on International Agricultural Research (CGIAR), about the potential dangers of the technology. The CGIAR Committee on Genetic and Policy Research, which met in Syria in April, decided to take urgent action to warn people about the dangers of the technology.

Realising the seriousness of the problem, the ICAR and the DBT set up a monitoring committee, which came out with a plan of action on May 19 to ensure that this lethal technology did not enter India. The Government of India refused licence to the patent holders to sell the Terminator seeds in the country. It also banned the import of such seeds and directed the quarantine office of the Customs Department, which has 18 stations spread over the country, to ensure that no transgenic seed material is permitted entry and that imported seeds get the approval of the Plant Protection Authority in the Ministry of Agriculture before they are released.

The ICAR-DBT committee has recommended an institutional mechanism, including a legislative instrument, to prevent the entry of Terminator seeds, especially through private seed companies. For this, the committee has recommended a single point of entry for imported seeds and such other materials. In the case of bulk imports, samples need to be drawn and analysed at ports and other points of entry to confirm that the material is free from the Terminator gene.

This would require facilities to check for transgenic material and quarantine suspect materials. Dr. Kuntala Jayaraman, Dean (Technology) and Director of the Department of

Biotechnology, Anna University, Chennai, regrets that the "quarantine offices are not competent enough to determine the quality of seeds." She suggested that laboratories be set up to test imported seeds and special training be given to quarantine and Customs officials.

According to Dr. Swaminathan, the quarantine arrangements at the various ports of entry in India are not adequate. The Customs Department's quarantine instructions are generally not adhered to. Travellers are supposed to declare food and agricultural items with the Quarantine Office. This rule, he says, has to be enforced strictly, and an awareness must be created about the potential dangers inherent in the current practice. Says Dr. Swaminathan: "Our Customs Department is oriented to look for gold, electronic items, computers and so on, not for seeds and plants." He suggests the setting up of separate units in the DBT for bio-safety, bio-ethics and bio-surveillance.

Close on the heels of the patent issued for Terminator has come the patent for "Verminator", granted to the British firm Zeneca. Verminator is also a chemically activated seed-killer. But unlike Terminator, which uses tetracyline, Verminator kills seeds by "switching on rodent fat genes that have been bio-engineered into crops". Dr. Swaminathan feels that Verminator technology is more sophisticated than Terminator. However, he says, both these patents expose the mindset of large seed companies, which are engaged in a trade war and are anxious to ensure that farmers buy seeds every year.

GMOs have kicked up a worldwide controversy. On June 7, Switzerland conducted a referendum on whether or not to permit the entry of GMOs. The U. S. is the only country where GMOs are extensively used. This is because the U. S. has installed bio-safety measures and effective quarantine facilities.

By Special Arrangement

It is possible that the Terminator technology will find application in all major food crops.

Defending Terminator technology, Dr. Harry. B. Collins, vice-president (technology transfer) of Delta and Pine Land, said at a Food and Agriculture Organisation meeting on food and genetic resources held in Rome in June: "The centuries-old practice of farmer-saved seed is a gross disadvantage to Third World farmers who inadvertently become locked into obsolete varieties because of their taking the 'easy road' and not planting newer, more productive varieties."

According to Dr. Hope Shand, RAFI's research director, such defence put out by the patent holders reveals a shocking ignorance about the way in which poor farmers conserve and use diversity, about the complexity of their farming systems, and about their economic realities and survival strategies in marginal farming environments.

According to Dr. Swaminathan, there is a need for a universal declaration on Plant Genome and Farmers' Rights similar to the universal declaration on Human Genome and Human Rights adopted by the United Nations Educational, Scientific and Cultural Organisation (UNESCO) in Paris on November 11, 1997. The UNESCO declaration stipulates in Article 10 that "no research or research applications concerning the human genome, in particular in the fields of biology, genetics and medicine, should prevail over respect for the human rights, fundamental freedom and human dignity of individuals."

According to Latha Rangan, research scholar at the Chennai-based M. S. Swaminathan Research Foundation, the Terminator gene is devised to imprint forever deprivation on the gene of the poor farmers. Says Dr. Hope Shand: "This is one way adopted by the developed countries not only to make a lot of money by selling seeds every season to farmers all over the world but also to call the shots on the agriculture and the livelihood and food security of the developing countries."

That is why it is important, says Dr. Swaminathan, not to stop just with strict regulations at the entry points, but to launch a large-scale public awareness and education programme.

Protection and Preservation

Indigenous communities all over the world have a wealth of knowledge of biological resources that enables them to derive a livelihood from nature. Yet diverse ecosystems and the traditional way of life of many rural communities are under the threat of extinction. Modernisation, industrialisation, monoculture in agriculture, the overuse of chemicals, and urbanisation have all contributed to endangering natural resources.

Plant resources are needed to produce food. The genetic diversity of these resources allows species to adapt to new environments, and survive new pests and changing climates. Seed banks, often referred to as gene banks, are an important tool in helping to preserve seed varieties from becoming extinct. Seed collections represent a significant share of existing agricultural heritage-the legacy of some 10,000 years of planting, ploughing and breeding of crops for human use. Collections of diverse crops are considered vital for developing innovations in the agricultural sector. Without them it may be only a matter of decades before agricultural yields fall and pests and disease take control of farmers' fields.

In the Jeypore Tract of Orissa in eastern India, the introduction of new high yield crop varieties (known as HYVs) together with forest degradation have led to a dramatic decline in the number of native plant species. To combat this threat and to ensure the security of their food supply, tribal communities from six villages, working with the MS Swaminathan Research Foundation, have initiated a programme to revitalise on-farm traditions. Through good quality seed production, farmer-to-farmer seed exchange systems, community seed banks and the establishment of community bio-diversity registers, local farmers are involved in participatory management of plant sources that enable them to secure an income while protecting natural resources.

4

Agricultural Bio-diversity in Asia

India is home to an immense diversity of both wild and cultivated crops. Yet the escalating rate of erosion of bio-diversity in India is causing great concern. It deprives millions of rural people, especially tribal farming communities, of their livelihoods. With the advent of the green revolution, the emphasis in agriculture has been largely shifted to the increase of yield-consequently a small number of rice varieties are being promoted, which have been selected for their capacity to give high yields through using high doses of fertilisers.

As a result, the genetic base of rice varieties is being reduced considerably and several traditional seed varieties are now facing extinction. These varieties, if conserved, will be of enormous value to agriculture by helping to improve food security and provide the genetic basis of new plant breeding programmes.

Plant genetic resources are important to farming communities not only for food production, but also for their role in culture and tradition. When plant diversity is lost, the genetic base of agriculture becomes depleted, as does a people's knowledge and cultural identity.

The loss of bio-diversity in agriculture can occur where crops are no longer being cultivated or are being marginalised. This is often referred to as the loss of genetic diversity. Replacing traditional varieties with high yield varieties can cause their extinction, and local knowledge of these varieties built up over centuries can easily be lost.

In Koraput District, local people are aware of the importance of conserving traditional rice varieties. In addition to the cultural value placed on inherited varieties, there are many benefits of traditional varieties over HYVs:

- superior taste and nutritive value;
- resistance to pests and diseases;
- resistance to droughts and floods;
- compatibility with local farming conditions;
- economic practicality-requiring fewer inputs such as chemicals and fertilisers;
- environmental sustainability; and
- tried, tested and proven performance.

Traditional Knowledge

Orissa was once the traditional home of the largest number of rice varieties of any state in India, more than 1,750. The number of local rice varieties is now only approximately 150. One of the biggest problems has been the introduction of high yielding and hybrid varieties of rice.

Together with non-availability of pure seed of different crops, absence of records of the genetic heritage in successive generations of farm families, and the consumption or destruction of seed during periods of food scarcity or displacement, the genetic resources of the area have dwindled.

Tribal farming communities in the area have played a significant role in preserving and conserving different traditional plant varieties. Tribal communities are knowledgeable about the kind of crops that can be cultivated in each type of land. Their use of the local resources and even their food habits reflect their knowledge of the environment.

Tribal communities in the Koraput District of Orissa have revitalised in-situ or on-farm conservation traditions through participatory community farming. These communities have been encouraged to support traditional varieties of rice and have established a seed bank to ensure food security and conserve bio-diversity.

Bio-diversity and Poverty Reduction

The tribal communities of Koraput District have used their knowledge and experience of forest species and agro-biodiversity to revitalise on-farm conservation by promoting traditional varieties of rice and medicinal plants. Community seed banks have ensured the preservation of bio-diversity and food security, enabling communities to obtain, store and manage vital seed. Many other important activities have helped conserve bio-diversity:

- The Protection of Plant Varieties and Farmers' Rights Act, recently passed by the Indian Government, recognises the role of farmers in cultivating and conserving bio-diversity. It legally provides for their recognition and monetary reward.
- Maintaining and managing the Community Bio-diversity Register, where information about flora and fauna is recorded, enables villagers to better understand the interaction between human, animal and plant support systems. The register has helped people to identify medicinal plants to help cure common diseases.
- A Participatory Plant Breeding Programme has introduced a local rice variety Veliyan that is good for fodder and soil conservation. It also produces a high yield of approximately 23 quintals per acre or 1 quintal per 176 square metres.
- Neighbouring villages have begun copying the process of conservation, evaluation and exploration of rice germplasms. Many NGOs from surrounding districts have adopted seed and grain banking and community medicinal plant gardens to enhance and conserve bio-diversity.

The conservation of medicinal plants has enabled people to understand the health and market value of plants in their region. As a result, some families are beginning to sell medicinal plants and forest products in the local market. Many other important activities have helped to reduce poverty, including:

- Organised exhibitions during festivals enable villagers to market tamarind, mohua, and gooseberry, for example, and receive good prices for these products.

- The availability of pure seed of local varieties of rice with high yields and improved fragrance has increased demand and encouraged some families to adopt organic farming methods. This has reduced the input costs to families used to the cultivation of HYV rice.
- Conservation of rainwater has led to community fish cultivation. The proceeds from the sale of fish go directly into a community fund, which is used for community-based micro-enterprises.
- Self-help groups have begun horticulture, poultry and fish breeding activities. All benefits from the activities go to the groups.

Community Seed Banks

Since agriculture began, farmers have kept seed from one growing season to the next, sometimes storing it for several years. Access to seed is essential to household food security where drought, crop failure and poverty erode plant resources. Community seed banks play an important role as seed reserves, giving access to planting materials. They help facilitate access to markets for farmers' produce, while seed fairs give farmers more choice over what they grow.

Storing *germplasm* in seed banks is inexpensive and requires little space. It successfully allows for the preservation of large populations with minimal genetic erosion. Seed banks also offer good sources of plant material for research. Through seed banks, researchers can obtain access to rare and endangered species without disturbing or damaging natural populations.

Seed Selection

Seed is collected from interested households in a specific quantity and cleaned by the women of the village. Seed is chosen according to the husk colour, grain size and other physical characteristics. Seed with impurities or mixtures of other varieties is not accepted. The seed quality is verified, purified, dried and monitored by both men and women members of the central village committee through traditional seed testing. Collected seed is then properly weighed and recorded.

Seed Storage

The tribal people in Koraput have their own indigenous way of storing crop seed and grains. Before preservation, seed is mixed up with dried powder leaves of Neem (*Azadirachta indica*) and Karanja (*Pongamia pinnata*), which helps keep seed free from pest attack.

Rice seed is stored in locally made structures called Dhoosi and Khaniki. The Dhoosi is made of long straw rope twined spirally. After the seed has been placed inside, the top is tied tightly. The Khaniki is a pot shaped bamboo basket plastered with cowdung paste. The cowdung paste acts as an insect repellent. After storing seed or grain it is covered with a bamboo plate plastered with cowdung paste. It is airtight, allows for minimum moisture content and does not get infested by insects. These structures are then stored in darkness within a community building. In general, seed bank storage usually follows a number of steps:

- collection;
- seed preparation;
- drying;
- packaging;
- storage;
- periodic germination testing;
- seed regeneration; and
- re-storage.

Documentation

The information required for proper documentation of banked plant seed includes: identification of seed variety and status, the location and nature of the population from which the sample was taken, details of the collection procedure, and information on storage conditions and viability.

Seed Distribution

The central village committee identifies families that urgently require seed. The secretary of the seed bank then records the variety and quantity of seed required by each family. Farmers receive seed on loan from the community seed bank, providing they agree to obey all the terms and conditions of the loan. On returning the loaned seed, each household has to give approximately 1.5 times the quantity taken.

The Future

The impact of the project has enabled rural tribal villages to become less dependent on HYVs and on outside inputs like fertiliser and pesticides. Productivity and farmer-to-farmer seed exchange have increased within the communities, in addition to growing independence from other villages. As a result, purified local seed varieties are available in each seed bank and farmers are using the optimum amount of seed for sowing. Women now use a modern seed and grain measuring and weighing system that protects them from being exploited.

The concept of community seed banks is being adopted by several other non-governmental organisations in communities within the district. Through these initiatives, remaining varieties of rice are being conserved and over-exploited medicinal plants are being cultivated in community gardens instead of being removed from the region's fragile forests. Market access allows communities to benefit financially from their conservation activities.

Cultivation Area of Agricultural Crops

Asian people grow and consume rice for the major food crop and thus about 90% of the world rice is produced in Asia. Percent cultivation area of rye and oats is much lower than the world average (6.1 and 6.8% of the world total). Sweet potato is also important crop in Asia. Among the horticultural crops, it is worthwhile to note that vegetables are quite popular among Asian people. The percent share of Asia in world vegetable cultivation area is 66.5%, which may be regarded as extremely high as compared to the world average or to those of other continents. The cultivation area of vegetables in Asia is 2.33 times larger than that of fruits. However, in most other countries, the cultivation area of fruits are considerably larger than vegetables (Lee et al., 1994). This inrger than vegetables (Lee et al., 1994). This indicates that vegetables are the major and most important horticultural crops in Asia not only in cultivation areas but also in other aspects such as life style, environment, economy, industry, research and education as described later in detail

That per capita production (or consumption) of horticultural crops in some Northeast Asian countries (Lee et al., 1994). It can be easily seen that vegetables are the major horticultural crops in Northeast Asian countries especially in China, Japan, Korea (North), Korea (South), and Taiwan. The ratios of vegetables to fruits are very high in these countries from 1.93 in Taiwan to 5.38 in Korea (S). This preference of vegetables among the Asian people has resulted in similar consequences in general horticultural areas; a large number of industries involved with vegetable seeds and seedlings production, fertilizer and agro-chemical production and distribution companies, storage and processing industries, and others. In addition, it is quite natural to find a larger number of research scientists working on vegetables in universities and colleges, governmental and private research institutes, and consequently a large number of research papers being presented at annual meetings of horticultural society and other related academic societies in many Asian countries. Research fund and budget also show similar trenrch fund and budget also show similar trends, in general.

Fermented and partially preserved vegetables are quite popular in Northeast Asian countries such as Japan, China, and Korea. For an example, the history of pickles in Japan is dated back to the pre-historic times, approximately 4,500 years ago. Many vegetables such as radish, Chinese cabbage, cucumber, and eggplant are pickled in many homes in Japan and other countries, but fairly a large amount of pickled vegetables are being traded between China, Thailand, Taiwan, and other nearby Asian countries (Inden et al., 1997).

Kimchi is a traditional fermented vegetable mix originated in Korea but now is being traded all over the world. Kimchi has its unique palatability giving sour, sweet, and carbonated taste. Approximately 300,000 ton of commercial kimchi products was produced by 337 processing factories in Korea in 1996

(Park, 1997) and export of kimchi has been increased drastically since last year. Hot pepper and garlic are vital ingredients for many kinds of kimchi. There are at least 187 different types of kimchi with distinctive recipes (Koo et al., 1991). These unique types of processed vegetables can now be found at any parts of the world from Australia to Canada (Nguyen et al., 1997; Wang, 1997) manufactured in commercial scale.

Production of Individual Horticultural Crops

Tomato, watermelon, cabbages, onion, and cucumber are 5 major vegetables in Asia and this coincides with the order in world total. Among the vegetables, garlic and eggplants are cultured mainly in Asia (87.3% and 90.3%), but carrot and green pea are not receiving much attention in Asia, as compared to other continents. Fruit data is expressed in production (1,000 ton) rather than cultivation area. Banana, including plantain, is by far the most abundant fruit in the world followed by oranges, grape, and apple. However, in Asia, apple is the first one followed by banana, mango, grape, and orange. There has been a tremendous increase in apple planted acreage in China and this will undoubtedly influence the apple industry in the East Asia and in the world in the near future (FAO, '997, Internet Search, fao.org/default.htm). Except for mango, the percentage is rather low in many fruits. The vegetable cultivation in Asia is indeed very advanced and characterized by the followings.

1. Various different cultivation types and techniques including numerous cultivars for each type.
2. Highly intensive cropping and frequent successive cropping in the suburban areas.
3. Well-advanced seedling growing techniques including grafting different types of herbaceous materials on commercial scale.
4. Considerable portions of the produces are being processed one portions of the produces are being processed or preserved in various types.

Vegetable and fruit production acreage in some Asian countries are China and India are the leading countries for vegetable production acreage followed by Turkey and Indonesia. However, fruit production acreage is highest in China followed by Turkey, Iran, and Indonesia. As far as production acreage is concerned, Asia has 66% of the world total vegetable production acreage as compared to only 44% of the world fruits production acreage. China has about 50% of the total Asian vegetable production acreage and 43.9% of the total Asian fruit production acreage, thus indicating that the Asian trade of horticultural products can be significantly influenced by only a few countries in Asia or by China alone! During the last 10 years, the protected cultivation acreage is China has been increased 17 times and the per capita consumption of vegetables also was almost doubled from 119.8 kg/year/person to 217.2 kg/year/person (Zhang, 1997). Average yield of vegetables per unit area was also more than doubled during the last 10 years. Introduction of promising, new hybrid cultivars may be responsible for this remarkable increases in yield per unit area. On the contrary, however, the use of hybrid seeds for vegetable production is extremely low in India, thus needing urgent attention and introduction of good hybrid cultivars (Agrawal, 1997).

Potatoes are the major vegetable crop in many Asian countries, not as vegetables in many circumstances but as an important food crop. Cauliflower, carrot, and eggplant are being cultivated only in some countries whereas cabbage, tomato, pumpkin, cucumber, onion, and watermelon are being cultivated in most Asian countries. Data base used in this report was mostly obtained from FAO Yearbook or internet FAO. However, even these data were not quite accurate and sometimes classified in different ways so that true data could not be easily accessible. Dried hot peppers are one of the important vegetable crops (or spice) and frequent international trades have been taken during the last 10 years. The data shows that the pepper producing area in India is more than 1.2 million ha. However, the FAO classification is very confusing and expressed inproperly so that more reliable as well as more accurate data should be compiled and available for precise evaluation and comparison of production statistics on a global basis. The data indicate again that more precise data is needed for possible accurate analysis of present production status and to set up goals for future guide for production and international trade.

International Trade of Horticultural Products

Due to the high population and limited arable land area, many Asians are experiencing food scarce and thus depend highly on imported foods from other continents or countries. As shown in grains account for more than 50% of the world trade followed by wheat and flour (23%), maize (8-9%), and rice (3%). Among the horticultural crops, banana, sometimes including plantains, is ranking number one in total amount of international export followed by oranges (including tangerines and mandarins), apple, potatoes and tomato. Pineapple (canned) is the only item which is mainly exported from Asia and most other horticultural crops are mostly exported from other continents. Asia does not import a large amount of horticultural crops, either. Major importer of horticultural crops is Europe, accounting for 50-80% of the total import for most horticultural crops. Potatoes and tomatoes are the major vegetables being traded worldwide.

Total amounts of export and import in Asian countries and some other countries. International trade of fruits and vegetables. International trade of fruits and vegetables in Asian countries and some other countries of the world can be obtained from FAO Homepage. Based on the total amount of export excess value, major exporting countries in Asia are Philippines, Israel, Syria, Turkey, and Myanmer, in that order. On the worldwide basis, however, Chile upsurged recently as the largest exporting country followed by Costa Rica, Ecuador, and New Zealand, in that order. The largest importing countries in the world are Canada and Russia, which are located in cold northern hemisphere. The largest importing countries in Asia are Israel, Hong Kong (now in China), Pakistan, Oman, and Philippines, in that order. However, the trade difference (export-import) is largest in Oman, followed by Hong Kong, Bahrain, Qator, and Pakistan. It is interesting to note that many of these importing countries are developing countries rather than developed ones, thus indicating that accumulation of high production technology rather than manpower is vital for producing and exporting "high quality" horticultural products.

Some Asian countries with large population, for example China, India, Indonesia, and Japan (71.7% of total Asian population in these 4 countries), do export and import horticultural products. However, the total amount and the differences between export and import are surprisingly low, thus meaning that production system for self-supplying that production system for self-supply is well stabilized in these countries. Even though vegetables are very much favoured by many Asian people, most of these horticultural products are being traded mostly among nearby countries with the exception of some fruits and ornamental plants. For example, Korea (S) imported a large amount of fresh and processed vegetables from China (Lee, 1994) and the amounts have been increasing steadily (FAO, 1996, 1998; Lee et al., 1994). On the contrary, Korea (S) exports considerable amount of fresh and processed vegetables such as pine mushroom, kimchi, dried mushroom, onion, and cucumber exclusively to Japan (FAO, 1998 Internet). Kimchi, which symbolizes the Korean traditional eating preferences because of its unique delicate favour and taste, is exported to all over the world, sometimes even under wrong name of 'Kimuchi', and now officially regarded as the well-known delicacies of the world and in every Olympic Games! In spite of the large amount of horticultural products, the amount of international trade has never been impressive until recently.

The bulk of horticultural crops compared to cereals or grains may be partially responsible for this. The maintenance of freshness in fresh vegetables and fruits would be possible only for short-distance transport. The dissimilarity in crops is another reason for this trend of international trade.

Furthermore, following regions or so called free trade following regions or so called free trade agreements (or areas) all over the world tended to aggravate the local trade tendency in some way or another. These are ASEAN (Association of South East Asian Nations, 9 countries), NAFTA (North American Free Trade Agreement, 3 countries), CERTA (Close Economic Relation Trade Agreement, 2 countries),

EU (European Union, 15 countries), LAIA (Latin American Integration Association, 9 countries), and many more agreements or areas such as Asian Pacific Region (APEC), East African Community, etc. The role and significance of ASEAN and AFTA (ASEAN Free Trade Area) have been drastically increasing until recently. However, financial crisis or ruin in some Asian countries such as Indonesia, Thailand, and Korea will greatly influence the ASEAN activities at least for 5-10 years. The role of OECD and APEC will be emphasized with relation to economies in Asian countries especially with the launching of WTO (World Trade Organization).

Professional Workers Engaged in Horticultural

Horticultural activities such as research, education, production, and industries are depending critically upon the number of horticulturists and specialists. In a given country, the number of registered members in any international horticultural society could be a good indication for the measure of global horticultural activities.

Expanding Horticultural Research

The Republic of Korea is probably one of the very few countries in the world with huge annual increases in governmental as well as Research foundation fund for horticultural research. The importance of horticultural products has been recognized in various different ways. For example, the percent share of the total amount of horticultural produce in all agricultural produces (including livestock and forest products) in 1995 was increased up to 24.4% for vegetables and 11.3% for fruit.

These values clearly indicate the importance of horticulture and horticultural industry. Similar trends have been maintained for last 5 years. The activity of the Korean Society for Horticultural Science (KSHS) members can be clearly confirmed by the number of registered KSHS members (+1,500), number of registered ASHS members and ISHS members, number of papers presented at the annual meeting of the KSHS, and the occupation of authors delivering the papers at the meeting. At the spring and fall meetings, several well-known scientists or horticulturists have been officially invited from other countries for special lectures. The KSHS also held several ISHS international symposia for the last few years and plans to host several more symposia in the next few years; on peppers, Post-harvest, and use of bio-regulators in fruits, etc. At the last ISHS international symposium on vegetable quality which was held in Seoul, Korea from October 27 to 30, 1997, we had 120 participants from 23 countries. Acta containing about 55 papers will be published in July this year (before the 1998 IHC in Belgium). The KSHS plans to celebrate its centennial anniversary of the official launching of modern horticultural research in Korea. The Agricultural Demonstration Station was established in 1906 in Suwon and, after several changes of its official name, has now become the Rural Development Administration (RDA) of the Korean Government, the centre for agricultural and horticultural researches in Korea.

The RDA now has National Horticultural Research Institute, Horticultural Sections in 3 Agricultural Experiment Stations, 9 Provincial RDAs, and 23 Commodity Experiment Stations mandated with horticultural crops. In addition to about 40 universities and colleges which have the department of horticulture, there are many private research institute, there are many private research institutes including 42 seed companies. Considering the total population of 45 million and only about 10% agricultural population (as of Dec 1997), the figures are rather too high as compared to most other developed countries. The export of horticultural produce is not impressive either except for some crops. For example, more than 70% of the grafted cacti being auctioned at Netherlands flower markets are exported from Korea. Korea also exports lilies and roses to Japan. Detailed data will be presented at the conference. Unfortunately, import has always well exceeded export as far as floricultural crops are concerned. With the booming of horticulture in Korea, the KSHS is very anxious to have the chance to become the host country for the 2006 IHC here in Korea rather than continual every other host in European countries.

Handling of Germ Plasm

Since vegetables are the major horticultural crops in Asia, considerable efforts have been focused on vegetable breeding and a large amount of vegetable seeds are being exported to Western countries. Brassica crops (cabbage, Chinese cabbage, broccoli and cauliflower) and radishes are some of the vegetables for export from Asia. Some seed companies in Japan, Korea (S), and perhaps Taiwan have been operating overseas companies or branch stations in other countries including the USA and also exporting countries including the USA and also exporting considerable amount of seeds. However, we expect some significant changes in seed industry because multinational seed companies such as Novatis and others began to operate official business in some Asian countries with their good seeds, advanced seed processing and cultivar identification technology in addition to the modern purity test methods.

Conclusions

With the rapid expansion of globalization under the influence of WTO, the world market for agricultural and horticultural produces has been increasing very rapidly and this, along with the introduction of noble cultivars with accurate identity so that the breeding efforts can be protected by the UPOV. However, such a rapid change does not likely take place due possibly to the very conservative attitude of Asian people, especially in their diet habit. In addition to good seeds of excellent traits, gradual and persuasive approaches are thought to be vital for the successful operation and maintenance of branch companies and persistent sales and distribution of seeds and other propagation materials. In spite of the sudden recent break on economic growth in some Asian countries, the role and significance of Asian horticulture will be persistently increased in the future, particularly in breeding, production, and utilization of horticultural products, especially of the vegetables. Horticulture in Asia will be of the vegetables. Horticulture in Asia will be continuing its steady progress and exerts significant role in world horticulture based on the following fact.

1. Asia has endless list of excellent germ plasm for various horticultural crops, especially for cruciferous crops including Brassicas, legumes, and native herbs (Zhu et al., 1997).
2. The number of Asians living in other continents of the world has been increasing drastically in recent years and it is more likely that they will certainly maintain their conservative palate (Kim et al., 1997; Sakiyama, 1997).
3. There are good markets developed, especially for processed horticultural products because of the recent economic growth in some Asian countries.
4. The preferences for vegetables and very intensive and labour-dependent culture will be maintained in most Asian countries.

Opportunities for the Food Processing Industry

In India agricultural and dairy sectors have achieved remarkable successes over the last three and a half decades. Besides being one of the world's largest producers of food-grains, India ranks second in the world in the production of fruits and vegetables, and first in milk production? providing much needed food security to the nation.

The accomplishments of the green and white revolutions have, however, not been matched by concurrent developments in supply chain management, and in new technologies for better processing, preservation, and storage of food. Pockets of shortages and near starvation, substantial wastages due to spoilage, quality deficiencies, and inadequate returns to the farmer are still very much in evidence.

Increased urbanization, improved standards of living, and the convenience needs of dual income families point to major market potentialities in the food processing and marketing sectors. This is also evident from the presence of several global foods giants and leading Indian industrial enterprises in the country's food processing sector, such as: Nestle India Ltd, Cadbury's India Ltd, Kelloggs India, Hindustan Lever Ltd., ITC-Agro, Godrej Foods and MTR Foods Ltd. Besides, in the current globalized milieu, our surplus food production, as well as the increasing preference for Indian foods (in several regions of the world) need to be leveraged to achieve economic, and strategic objectives through exports. The Food and Agriculture Integrated Development Action (FAIDA) report (1997) prepared by McKinsey has estimated that, driven by changing consumer preferences, the annual consumption of 'value-added' foods alone would grow to Rs.225, 000 crores by 2007? larger than the entire manufacturing sector! A more recent report has stated an absolute revenue increase of Rs. 900 billion in food manufacturing between 1993 and 2000. This is in contrast with Rs. 150 billion and Rs. 300 billion in the pharmaceutical and IT industries, respectively. Overall, the value of the Indian food industry has increased from Rs. 3.09 trillion in 1993-94 to Rs. 3.99 trillion in 2000-01. The segments with the largest growth potential have been identified as dairy, wheat, fruits and vegetables, and poultry. This report has also identified some of the major challenges for the emerging food industry in India.

5

Mediterranean Fruit Crops

The current situation of nine temperate Mediterranean fruits and nuts in the world, will be discussed in descending order of economic importance.

Olive

The cultivated olive (Olea europaea, Oleaceae) is a long-lived, evergreen tree native to the Mediterranean basin. It is valued for its fruit and oil. Mediterranean countries account for around 95% of the world's olive cultivation (8,702,000 ha). In the Mediterranean 90% of the olive trees are grown for the oil. World olive oil production is currently increasing; the commercial crop during 1990-1994 was 1,796,900 t (IOOC 1994), with the European Community (EC) producing around 1,337,000 t (Spain 40%, Italy 33%, and Greece 22%), Tunisia 194,000 t and Turkey 61,500 t of oil. World table olive production in 1990-1994 was 954,500 t (IOOC 1994). Spain is the world's largest producer (23%) and exporter of table olives, follow by Turkey (12%), U. S. (11.4%), Morocco (8.5 %), Syria (7.5%), and Greece (7.3%). In Australia, Chile, China, Mexico, New Zealand, and South Africa, olives are considered a new crop.

The olive oil and table olive industries play an important role in the agricultural and processing sectors of the major olive producing countries. Most olive oil is consumed within the Mediterranean countries; only 18% of production enters world trade. On average from 1988 through 1991 the world olive oil marketed represented 6% of the quantity and 23% of the value of the world trade in fluid edible oils (UNCTD 1993). To a large extent olive oil does not compete with other vegetable oils but occupies a speciality niche market.

The olive tree has a wide range of adaptability. It requires a mild climate with warm summers and cold winters. The tree requires substantial chilling for good fruiting (Martin et al. 1994), but is injured when temperatures fall below-10°C. Olive is considered a drought-resistant species because it thrives in areas where water stress is frequent: Mediterranean climates. It has been postulated that the minimum water requirement for olive is 2,000 m3/ha year, mainly during flowering and fruit setting in late spring, and again in the summer as the fruit increases in size (Tous 1990; Bongi and Palliotti 1994). Olive trees will grow on poor soils and rocky hillsides, but deep soils produce the best quality fruit. They tolerate saline or alkaline soils and those with a high lime content. Their root system is relatively shallow, and will not tolerate waterlogged soils.

Olives are wind pollinated, thus flowering during rain, high temperatures and dry wind conditions are deleterious to good fruit set (Griggs et al. 1975). Mature olive trees produce huge numbers of flowers, but the fruit set is normally below 5% (Lavee 1986; Martin et al. 1994). Most olive cultivars will set some fruit in a monocultivar culture. However, they benefit greatly from cross-pollination (Lavee 1986). Olives are picked late in autumn or winter, as the oil content and fruit characteristics change with ripening. Olive cultivars usually fall into one of two commercial uses (Jacoboni and Fontanazza 1981; Barranco and Rallo 1984; Tous and Romero 1993): "Oil" and "Table." "Oil" cultivars predominate. The most famous are: 'Picual', 'Arbequina', 'Cornicabra', 'Hojiblanca', and 'Empeltre' in Spain; 'Frantoio', 'Moraiolo', 'Leccino', and 'Pendolino' in Italy; 'Koroneiki' in Greece; 'Chemlali' in Tunisia; 'Ayvalik' in Turkey; 'Mission' in California and Australia, etc. The "table olive" cultivars include 'Manzanilla' and 'Gordal' (syn. 'Sevillano' in California and 'Queen' in Australia) from Spain; 'Kalamata' from Greece; 'Ascolano' from Italy; 'Barouni' from Tunisia, etc. Alternate bearing is a problem with some cultivars.

Tree management of this crop has undergone many changes in recent decades including increased planting densities (100 to 300 trees/ha), new methods of propagation, improved cultural practices and mechanical harvesting, are some examples.

Growth regulators, such as naphthaleneacetic acid (NAA), are used for fruit thinning to increase the fruit size and control the alternate bearing in some table olive orchards (Lavee 1986; Martin et al. 1994). Olives are propagated primarily by cuttings, truncheons or by budding seedling rootstocks. The trees propagated by rooting semi-hardwood cuttings come into bearing within three to four years after planting. The type of harvest depends upon fruit use. Oil olives are harvested by hand or by mechanically shaking the tree. Those for pickling must still be harvested by hand as the fruit bruises easily. The major pest problems are the following: olive fly (Dacus oleae, mainly in the Mediterranean basin), black scale (Saissetia oleae), olive leaf spot (Spilocaea oleaginea), olive knot (Pseudomonas savastanoi), and Verticillium dahliae.

Olives are nutritious, the oil component includes unsaturated fatty acid (70% to 80% oleic acid and 7% to 12% linoleic acid), and small amounts of polyphenols, tocopherols, sterols, and many aromatic compounds (Mataix and Martinez 1988; Tous and Romero 1993).

Olive oil and table olive industries play an important role in the agricultural and processing sectors of the major olive producing countries of the Mediterranean basin. World olive oil production and exports are projected to increase slightly in the next few years, while table olive production will remain stable. While Italian olive oil production is expected to stabilize or decrease, Spain, Tunisia, Morocco, and other countries (Argentina, Australia) are expected to achieve substantial increases in this crop (UNCTD 1993). In addition to culinary uses, olive oil, an unsaturated fat, has recently become more valued for its health benefits. Mainly for this reason, through the last decade the olive oil consumption has increased in several non-Mediterranean countries (USA, Australia, Japan, etc.).

Mandarin

The mandarin (Citrus reticulata, Rutaceae) is the largest and most varied group of edible citrus. It is a native of China and Southeast Asia, and was known and extensively planted in China and Japan in the 16th century, but was not introduced to Europe and North America until the 19th century (Webber 1967; Saunt 1990). World production of mandarins is currently increasing; the commercial crop in 1993 was 9,570,000 t, 12.5% of the total world citrus production (FAO 1993). Spain and Japan each account for 16% of the world production. Other major producers in descending order of production are China (8%), Brazil (7%), Korea (6.5%), Italy (5%), United States (4.7%), and Morocco (4%). While production by Japan has decreased around 40% over the last decade, that of Mediterranean countries has increased by 50% (Saunt 1990; Ferguson 1992).

The mandarin has a wide range of adaptability and is grown under desert, semi-tropical and sub-tropical Mediterranean climatic conditions. Mandarin trees and their hybrids are usually the most cold-resistant of all commercially grown citrus. However, mandarin fruits suffer more frost damage than most oranges and grapefruit. The climate in which mandarins are grown has a great influence on fruit quality. Most commercial mandarins are irrigated, but excessively wet conditions should be avoided as this may lead to increased disease. Windbreaks are commonly planted to protect crops from high winds. Light, fertile loams are ideal, although a wide range of soils will produce good crops.

Mandarins are called "slip-skin" oranges "zipper skin" or "easypeeler" because they have loose skins and are easy to peel and separate into segments. The fruit has long been appreciated for its distinctive and sweet flavour and aroma. They are used primarily for eating out-of-hand, as fruit sections, in fresh juice, and to a limited extend for processing (Young 1986). Mandarin cultivars usually fall into one of three commercial groups (Bono et al. 1985; Saunt 1990): "Satsumas" ('Owari', 'Okitsu', and 'Clausellina'); "Clementine" ('Arrufatina', 'Clemenules', 'Esbal', 'Fina', 'Hernandina', 'Marisol', and 'Oroval'); and "other mandarins and hybrids" ('Dancy', 'Fairchild', 'Fortune', 'Kara', 'Nova', and 'Wilking'). Careful selection of three or four mandarin cultivars per orchard can yield fresh fruit for an eighth-month ripening period, beginning in mid-Sept. with 'Okitsu' and 'Clausellina' and extending into

Apr.-May with late cultivars such as 'Fortune'. Some cultivars such as 'Comun mandarin', 'Dancy', 'Fairchild', 'Fortune', require cross-pollination for best fruit production (Eti and Stosser 1992).

This evergreen tree requires close plantings (2 x 5 m) and minimum heights for easier harvesting; in this dense planting system it is customary to hedge prune mechanically. Growth regulators are used to control alternate bearing and substitute for pollination, which produces excessive seeds. Mandarins are propagated by budding selected scions on to nucellar seedling rootstocks; a commonly used rootstock is 'Cleopatra' mandarin (Samson 1980). However, any of the rootstocks noted for producing high quality fruit are suitable. The tree comes into bearing three to four years after planting. Most of the pests that attack citrus may be present in mandarin orchards. The major pests are aphids, scale insects, whiteflies, citrus thrips, fruit flies and mites.

Overall, world mandarin production and exports are projected to lightly increase in the next few years. While Japanese production is expected to stabilize or decrease, Spain, Morocco, and other Mediterranean countries are expected to achieve substantial increases in this crop. The excellent quality and characteristic flavour of the new mandarin cultivars are highly prized by some, and if seedless varieties of larger size can be obtained, their popularity will greatly increase.

Fig

The fig (Ficus carica, Moraceae) probably originated in Western Asia, and spread to the Mediterranean. Today, the fig is a moderately important world crop, with an estimated annual production of one million tons of fruit (Sadhu 1990). Approximately 40% of this crop is sold as dried fruit. About 30% of the crop is produced by Turkey (300,000 t). Other major producers in descending order are Egypt (160,000 t), Morocco (57,000 mt), Spain (50,000 t), Greece (50,000 mt), California (43,000 t), Italy (40,000 t), Algeria (38,400 t), Syria (37,000 t), Tunisia (35,000 t), and Libya, Iraq, and Portugal (Sadhu 1990; CIHEAM 1994). While production by Italy and Spain has decreased over the last decade, that of Turkey, Syria, and Brazil has increased (IBPGR 1986).

The fruit usually is consumed fresh locally or in dried, canned, and preserved form. Dried figs and those unfit for human consumption, can be used as animal fodder. Several countries import dried figs or the paste. The main exporters of dried figs and paste are Turkey and the United States. Of California's production, 85% is marketed as dried figs, 12% as canned figs and fig juice, and 3% as fresh fruit (Storey 1975). The nutritional value of fresh figs is comparable to that of many other fruits. They are high in calcium. Dried figs, with only 20% water are nutritious relative to other fresh fruits.

The typical fig-producing regions have mild winters and hot dry summers. The fig tree is tolerant of a wide range of environmental conditions, has a low chilling requirement, will withstand some frost and is tolerant of drought, although it grows most vigorously with abundant water. A frost of-5° to -10°C may kill the plant down to ground level. Figs can be grown on a wide range of soils, including heavy clays, loams, and light sands, but ideally the soil should be well-drained. The plant is moderately tolerant of high salinity (IBPGR 1986).

There are two main commercial types of figs, the "common fig" that produces fruit without pollination, and the "Smyrna fig" that requires pollination by a fig wasp (Blastophaga spp.), that lives in the "caprifig" (male fig), to set fruit. The "common-type" (self-pollinated) fig is more commonly grown. These cultivars bear one or two crops per year. The most famous are: 'White Adriatic' (syn. 'Verdone'), 'Black Mission', 'Kadota' (syn. 'Dottato' in Italy), and 'Conadria' in California (Storey 1975; Ferguson et al. 1990); 'Cuello de Dama Blanco' and 'Napolitana Negra' in Spain (Llacer et al. 1994); 'Kalamon' in Greece (Lionakis 1994); 'Sultani' in Egypt and Tunisia (Mansour 1994; Mars 1994); 'Rhouddane' in Morocco (Loudyi 1994), etc. The "Smyrna-types" include the popular Turkish cultivar known as 'Sarilop' in Turkey and 'Calimyrna' in the United States and other cultivars of Algeria 'Taranimt' and 'Tameriout', that have high dried fruit quality (Rebour 1955; Aksoy 1994).

The tree is propagated by rooting 20 cm cuttings of one to three year old wood taken during the dormant season. Trees are normally planted 4 m apart with 5-6 m between rows.

Fruiting begins after three years. Regular fertilization will increase yields without reducing fruit quality. Figs carry two crops in a year, the main crop normally being the second crop in late summer and autumn. The first, or "breba" crop, is produced from flowers initiated in the preceding late summer and maturing from May to June. The second crop is produced from flowers on the current season's growth. Pruning may be required to maintain a balance between new and old wood, as well as to remove suckers and to keep the tree's canopy to a reasonable size for easy harvesting.

Fresh figs are picked when they begin to soften and the colour change indicates maturity. When picking, gloves should be worn to prevent damaging the fresh fruit and to prevent the skin irritation caused by the white sap that contains ficin exuding from the broken stem. Since fresh figs ripen irregularly, picking should be done daily or weekly during the long harvest period (4-6 weeks). In California most figs are grown for drying. They are mechanically harvested by sweepers from the ground during Sept. and Oct. (Obenauf et al. 1978). After harvest, the dried figs are washed and can be stored for a few days at 0° to 1°C. Fruit is dried in the sun or by using an electric dryer at a temperature of 60° to 70°C before processing as dried figs.

Figs are not usually seriously affected by pests except in high rainfall areas. In these areas and during the rainy season, fruit cracking usually occurs and fungicide sprays may be necessary to control surface rot (Alternaria alternata), smut (Aspergillus niger), and mold of fig (Botrytis spp., Penicillium spp.). Aphids, birds, fruit flies, and scale insects are occasionally a problem. Figs are highly susceptible to nematodes and should not be planted in infested soils.

The economic importance of fig production is likely to continue into the future. In the world market, there is an increasing demand for fresh figs and a stable demand for dried figs. The most important trade aspects of this species are the short commercial life of the fresh fruits, and for the dried fruits, the market competition of Turkish production, where production costs are lower than for other countries (Europe). Fig production in some European countries is slowly decreasing. Turkey and

Mexico are expanding their fig production. At present, evaluation of fresh cultivars in Europe and 'Calimyrna' improvement in the United States, combined with improved cultural practices and better fresh fruit Post-harvest practices are opening new prospects for this crop.

Persimmon

The persimmon (Diospyros kaki, Ebenaceae) is native to East Asia, most likely China. It has been cultivated in the Mediterranean basin for more than a century (Morton 1987). Europeans refer to the persimmon by the Japanese name kaki. World production is about 1,200,000 t annually from an area of approximately 235,000 ha, with China producing around 57% and Japan 27% of the crop. Smaller, but expanding persimmon industries are being developed in Italy (70,000 t), Korea (50,000 t), Brazil (45,000 t), Israel (10,000 t), Spain (8,000 t), the U. S. (7,000 t), New Zealand, and Australia (George and Nissen 1990; CIHEAM 1994; Mowat and George 1994). Persimmons are a deciduous, monoecious plant, and set fruit parthenocarpically or with pollination. They grow well in warm-temperate and subtropical areas, and tolerate the hot and humid coastal areas. They can tolerate winter temperatures of -12°C but require no winter chilling. Persimmons are grown on a wide range of soils but produce best on deep, fertile, medium-textured soil with good drainage. They produce better than most fruit trees on heavy clay soils. Ferguson and Arpaia (1990) showed that, under California San Joaquin Valley conditions (145-367 mm), they require at least 9,000-12,000 m3/ha per year of irrigation in addition to annual rainfall.

The fruit is oblong-conical and shows wide variation in size, shape and colour. Fruit may be yellow, orange, or red when ripe. Fruit is normally harvested from late Sept. to early Dec. for the fresh market or for drying. The fruit is sweet but most cultivars have an astringent taste until fully ripe. The cultivars are broadly classified for horticultural purposes into two major groups (Itoo 1980; Kitagawa and Glucina 1984; Morton 1987): "astringent" and "non-astringent," based on their flavour at maturity. The most important cultivars are 'Hiratanenashi', 'Hachiya', 'Tanenashi', 'Triumph', 'Kaki Tipo', and 'Rojo Brillante' (astringent); 'Fuyu', 'Jiro', 'Hana-Fuyu', and 'Gosho' (non-astringent). Until recently, most cultivars were commercial the astringent type. These can be treated with ethylene, CO_2 and other chemicals to remove the astringency while the fruit is still firm (Itoo 1980; LaRue et al. 1982).

The trees are trained to a modified central leader and require 7-10 years to reach full bearing. Propagation is by budding or grafting onto seedling rootstocks. Persimmons are relatively free of pests and diseases. The production problems depend on the growing areas. In Mediterranean countries the most pressing pest is the fruit fly (Ceratitis capitata), while in others (Japan, California, etc.) fungi such as anthracnose (Gloesporium kaki) and grey mold (Botrytis cinerea) are common diseases. Physiological problems include cracks radiating from the calyx, sunburn, post-flowering fruit drop, and alternate bearing.

The kaki or Japanese persimmon is mainly eaten fresh, but can be frozen, canned or dried, and is sometimes used in oriental cooking. They can be stored for up to 6 months in modified or controlled atmospheres. Persimmons are high in vitamin A, and are a moderate source of ascorbic acid. Some astringent cultivars contain an average of 1.4% soluble tannins in skin cells when not fully ripe.

The persimmon has long been appreciated in the Far East as China and Japan together produce 80% of the world crop. The overall production trend is stationary. While Japanese production is expected to decrease, Brazil, Israel, Spain, Australia, and New Zealand are expected to increase plantings of this crop. The future of this fruit depends on the selection of high quality cultivars and specific marketing techniques to ensure a firm non-astringent persimmon reaching the consumer (George and Nissen 1990).

Pomegranate

The pomegranate (Punica granatum, Punicaceae), native to Persia, is one of the oldest known edible fruits. The total world crop was estimated at 800,000 t of fruit (IBPGR 1986). The tree has been cultivated extensively in the Mediterranean countries (Tunisia, Turkey, Egypt, Spain, and Morocco), Iran, Afghanistan, India, and to some extent in the United States (California), China, Japan, and Russia. The fruits normally are sold in the local markets of these countries, except Spain and Turkey which also export. The crop requires a long hot summer to mature the fruit, yet can withstand low temperatures of -9° to -12°C in the winter and is extremely drought-tolerant. Pomegranates grow naturally as a bushy shrub or a small tree with spiny branches, in subtropical areas with arid summers. The pomegranate bears heavily and regularly every year, and requires very little care. The tree adapts to a wide variety of soils.

The fruit is a false berry, subglobose in shape, shiny red, yellowish green or whitish, crowned with a persistent calyx, and has a hard rind. Internally, the fruit has few to many locules containing many seeds. The testa of the seed is filled with a whitish pulp (aril) and an acid juice. The seeds may vary in size and hardness. Fruits with hard seeds possess poor eating quality and, therefore, cultivars producing soft seeded fruits are preferred. Hard seeded fruits, having higher juice content and an intense red colour are utilized for processing (Kumar 1990). Several cultivars are grown commercially in the world: 'Wonderful' in California and Israel; 'Mollar' and 'Tendral' in Spain; 'Schahvar' and 'Robab' in Iran; 'Hicaz' in Turkey; 'Zehri' and 'Gabsi' in Tunisia; 'Alandi' and 'Ganesh' in India, etc. (LaRue 1980; Morton 1987; Patil and Karale 1990; Aksoy 1994; Llacer et al. 1994; Mars 1994). The pomegranate is both self-pollinated and cross-pollinated by insects. The fruit is consumed fresh or can be processed into juice, a syrup (grenadine), jams, or a type of wine. The juice content of pomegranate fruit accounts for about 45% to 65% of the whole fruit or 76% to 85% of the arils. Lists general compositional data of pomegranate pulp. Pomegranates are also valued for their therapeutic properties (Kumar 1990).

The tree requires minimal pruning, except for removing suckers at the base of the plant and giving some shape to the canopy. It is propagated mainly by hardwood or softwood cuttings. The tree comes into bearing three to four years after planting. The fruit is harvested by hand from mid-Sept. to early Nov. The most important problems cause cracking of fruits that affects the pomegranate quality in storage. In India and Egypt the fruit borer (Virachola isocrates) is a serious pest (IBPGR 1986; Patil and Karale 1990). Different Post-harvest diseases such as Aspergillus niger, Botrytis cinerea, and Alternaria sp. cause the fruit to split before ripening. The practice of picking early in the season to avoid fruit cracking is known to result in poor quality fruit (Kumar 1990). The pomegranate is resistant to the oak root fungus, Armillaria mellea.

Crop production is stable. Turkey and India may increase the area planted to pomegranate. The future of this fruit depends on the selection of high quality cultivars with soft seeds, higher total soluble solids and richer colour (Kumar 1990) and fruits resistant to cracking and fruit borers.

Pistachio

The pistachio, (Pistacia vera, Anacardiaceae), probably originated in Central Asia, though it has often been referred in early histories of Western Asia, Asia Minor, and the Mediterranean countries. The fruits are classified as drupes with edible seeds. World production of pistachios is currently increasing, and the commercial crop in 1993/94 was about 258,000 t in-shell (INC 1995). The main producers are Iran (43%), the United States (26.5%, Sacramento and San Joaquin Valleys in California), Turkey (19.5%), Syria (8.5%), Greece (1.5%), and Italy (1%). Lesser quantities are produced in Lebanon, Tunisia, and Australia. Spain is a big importer of pistachio nuts, almost 10,000 t in-shell (Vargas 1990). There are good market possibilities for expansion of the pistachio crop in Turkey, Spain, and California (Chang 1990; Galan-Sauco and Rallo 1993).

The native range of pistachio is characterized by long, hot, dry summers and moderately cool or cold winters. In California the tree is grown in areas where winter temperatures accumulate 1,000 h under 7°C, the conditions required to break the rest period for normal growth and fruiting of 'Kerman' pistachio (Crane and Maranto 1988). Although pistachios will grow in rocky shallow soils, productivity is enhanced in a deep well-drained soils and by proper cultural practices. They tolerate saline or alkaline soils and those with a high lime content. Though originally from a xerophytic environment, they require more water than most fruit trees to produce well, approximately 10,000 m^3/ha per year under California Central Valley conditions (Ferguson and Arpaia 1990). In other areas of the Mediterranean basin, such as Southern Italy, the recommended minimal irrigation of 500-600 m^3/ha/year in the summer maintains good tree performance (Caruso 1990).

The pistachio is a moderately sized deciduous and dioecious tree (3 to 8 m). In commercial orchards the ratio of staminate trees to pistillate trees is 1 to 8. Pistachios have an extended juvenile period and require 5 years of training to establish a modified central leader canopy and 7 to 10 years to reach full bearing under good irrigated conditions. It is also a strongly biennial-bearing species, as fruit embryo growth induces the abscission of already induced floral buds (Crane and Iwakiri 1981). For example in Turkey the ratio of "on" to "off" trees is 1:1 (Kaska 1994). Cultivars differ among countries and are budded on seedlings of different Pistacia rootstocks. California's industry consists of one pistillate cultivar, 'Kerman', one staminate cultivar ('Peters') and four rootstocks (P. terebinthus, P. atlantica, P. integerrima, and the hybrid P. atlantica x P. integerrima). 'Kerman' was imported from Iran (then Persia) and named here for its province of origin and Peters for the Fresno grower who selected it. Other cultivars growing in others countries are 'Momtaz', 'Owhadi', and 'Kalehghouchi' in Iran; 'Uzun' and 'Kirmizi' and Turkey; 'Red Aleppo' in Syria; 'Aegina' in Greece; 'Bianca' in Italy; 'Mateur' in Tunisia; 'Larnaka' in Cyprus; and 'Sirora' in Australia; (Vargas 1985; Crane and Maranto 1988; Vargas et al. 1994).

The nuts are borne in clusters that can be harvested by hand, by knocking with a pole or by shaking. The crop matures in early autumn. Mechanical harvesting and processing procedures have been well developed in California. Aflatoxin contamination by the fungus Aspergillus is generally not a problem in California, as all pistachios are processed within 24 h of harvest (Ferguson and Arpaia 1990). No serious diseases of pistachio have been reported in Turkey or Iran. In California, however, many orchards have been seriously threatened by Verticillium wilt (Verticillium dahliae). Recent orchard surveys in California have revealed other diseases, such as Botrytis blossom and shoot blight (Botrytis cinerea), and Botryosphaeria panicle and shoot blight caused by Botryosphaeria dothidea are increasing in California (Michailides et al. 1988).

Pistachio nuts may be bought roasted and salted, shelled or unshelled, and are used in ice creams, sweets, cakes, and meat dishes. The kernels are nutritious, with a fat component including 69% oleic acid and 19.8% linoleic acid (Woodroof 1979). Pistachio nut oils are also used in the cosmetic and pharmaceuticals industries (Caruso 1990).

World production of pistachios is currently increasing. The United States (California), Iran and several countries of the Mediterranean basin (Turkey, Greece) are expected to increase their planted hectares. The most serious problems of this tree are alternate-bearing, the soil borne fungus Verticillium dahliae, and the long juvenility.

Carob

The carob (Ceratonia siliqua, Fagaceae) is a perennial leguminous tree native to the Mediterranean basin and southwestern Asia (Smith 1976). Its cultivation began before written history. This species is also known as "St. John's Bread," "algarrobo" in Spain, "carrubo" in Italy, and "caroubier" in France. The beans and kernels are useful for a variety of purposes including food, fodder, and as a source of industrial products such as gums, sugar, and alcohol. The tree is also used for shade and erosion control (NAS 1979). World production of carob pods was estimated about at 330,000 t (Tous and Batlle 1990), mainly in Spain (45%), Italy (16%), Portugal (9%), Morocco (7.5%), Cyprus (6%), Greece (5%), and Turkey (4.5%). Lesser quantities are produced in Algeria, Tunisia, and Israel. Spain is also the world's largest exporter of carob beans. The only significant recent plantings outside the Mediterranean region are in Australia (Esbenshade and Wilson 1986; Tous 1995).

The tree grows best in a Mediterranean-type climate with cool, not cold, winters, mild to warm springs, and warm to hot summers with little rain. A frost of -4°C or below may kill the young trees, or shoots and flowers in mature trees. Although drought-resistant, carob trees do not bear commercial crops unless they get enough rainfall, a minimum average rainfall of about 500-550 mm is needed to ensure a profitable crop (NAS 1979). The tree survives in a wide variety of soils, like pistachio, but must have good drainage. Although it is a legume, it is unable to fix nitrogen (Martins-Loucao 1985).

The carob is a large (7 to 10 m) trioecious tree with staminate, pistillate and hermaphrodite inflorescences in different trees. Pollination is by insects (bees, flies, etc.) and the wind (Tous and Batlle 1990). In new commercial orchards the ratio of staminate or hermaphrodite trees to pistillate trees should be 1:8. In traditional orchards male branches are grafted on female trees to provide pollen in situ. The carob, like the pistachio, matures slowly and most budded cultivars begin to bear in the 5-6th year after planting. It is also a strongly alternate-bearing species, particularly when growing conditions are poor. Today, commercial cultivars differ in pulp quantity or yield of seeds. The most important cultivars are: 'Negra', 'Rojal', 'Matalafera', and 'Duraio' in Spain; 'Gibiliana' and 'Amele' in Italy; 'Mulata' and 'Galhosa' in Portugal; 'Tylliria' in Cyprus; 'Hemere' in Greece; 'Sfax' in Tunisia; and 'Santa Fe' in California (Tous and Batlle 1990).

The tree requires little pruning, and minimal cultural care. Carobs respond to N fertilization by increasing pod yield (Lloveras and Tous 1992). They are propagated mainly by budding or grafting onto seedling rootstocks. The flowers, which are borne in small lateral racemes or clusters, bloom during Sept.-Nov. on older wood. The carob bean is a dark-brown, fleshy edible pod, 10-30 cm long, containing 5-15 hard seeds, and ripening in late summer, during bloom. Carob seeds comprise about 10% to 20% of the weight of the pod. Fruits can be harvested by hand, by knocking down with a long pole, or by shaking. No serious problems of carob have been reported in the Mediterranean area. The major pests are the stem borer

(Zeuzera pyrina) and carob moth (Myelois ceratoniae), and only one fungus, Oidium ceratoniae, is occasionally a problem in the leaves and young pods.

The carob pod is very rich in sugar (up to 45%) and low in protein. Mixed with other feeds higher in protein, it is good source of stock feed (horses, calves, goats, milk cow). Finely-ground pods make a sweet nutritious human food, and are used in a similar manner to cocoa in cakes, drinks, puddings, breakfast cereals, and candy. The carob powder has more vitamins, minerals, and fewer fats, carbohydrates, and calories than chocolate made with cocoa. Additionally carob powder contains no caffeine and theobromine (Whitheside 1981). The carob gum or locust bean gum (LBG) surrounding the endosperm is an important commercial stabilizer and thickener in many food industries (bakery goods, ice creams, baby foods, sauces, jams, cheese, jellies, and canned meats), as well as in pet foods, cosmetics, and pharmaceuticals (NAS 1979; Morton 1987; Di Lorenzo 1990; Tous and Batlle 1990).

The overall production trend is stationary. Several countries, such as Spain, Portugal, and Australia, are expected to have new plantings of this crop. In the world market, there is an increasing demand for carob seed and a stable demand for carob pulp. In the last 5 to 10 years, because of the rising prices of the carob seed derivatives (LBG), the carob has been the object of renewed interest (Tous and Batlle 1990).

Cactus Pear

The cactus pear (Opuntia ficus-indica; Opuntia spp., Cactaceae) is a fleshy bush or small tree, native to the desert zones of Northwestern Mexico and the Southwestern United States (Knight 1980). The plant was brought to Europe by the first Spanish colonists from Mexico and has been cultivated along the Mediterranean coast since the late 17th century (Retamal et al. 1987). This fruit is also known as "barbary fig," "tuna," and "Indian fig." The total world crop was estimated at 300,000 t of fruit. Mexico's estimated current production of 200,000 t of fruit on an area of 50,000 ha (Flores and Gallegos 1994). The other important production region is the Mediterranean basin (Italy, Spain, Egypt, Morocco, and Israel) and to a more limited extent South America, South Africa, Southwestern Asia, the U. S., and Australia. The fruits normally are sold in the local markets of these countries, except for Mexico, Italy, and Chile which are also exporters. The cactus pear is not cultivated as a regular commercial crop, but rather is planted as fences, windbreaks, and gardens. The only significant plantings are in Mexico, Italy, Spain, Chile, Israel, and South Africa. The plant may reach a height of 5 m and prefers calcareous soils and semi-arid climate with temperatures ranging from 18° to 26°C.

The Opuntia is a cactus in which the branches or stems ("cladodes" or "nopalitos") have the appearance and function of a leaf. The fruit is generally pear-shaped and has a number of small spines, with a soft juicy pulp that contains numerous small black seeds. Fruits ripen and are harvested by hand from July to Dec.

The plant requires minimal cultural care. It is propagated mainly by vegetative means (cladodes or shoots). Fruiting begins at 2 or 3 years and reaches full production around 7 or 8 years. A mature plant can yield 100 to 200 fruits (Knight 1980). Many local cultivars have been developed, and generally are named for their fruit colour at maturity, particularly in Mexico ('Reyna', 'Rojo Pelona', 'Esmeralda', etc.), Italy ('Gialla', 'Rossa', and 'Bianca') and Spain ('Verdales', 'Morados', 'Sanguinos', and 'Blancos') (Barbera et al. 1992; Flores and Gallegos 1994; Llacer et al. 1994). Future demand for this fruit depends upon developing high quality cultivars without spines and few seeds. Due to its xerophitic habitat the cactus pear does not have serious problems with pests and diseases. However in some areas the fruit fly (Ceratitis capitata) causes damage to the fruits, particularly during years with mild weather during autumn (Barbera et al. 1992; Lionakis 1994).

The vegetable stems ('nopalitos') and fruits of cactus pear are useful for a variety of purposes including food (fresh fruit, paste, jam, salads, and refreshing drinks), fodder (auxiliary feed for cattle, sheep, and goats), medicinal (antidiabetic agent) and for industrial products such as alcohol, soap, pigments, pectins, and oils (Lakshmi Narayana 1980; Borrego et al. 1990;

Hegwood 1990). The present trend of the production of this fruit is increasing. In the future, Mexico, Italy, and South Africa are expected to increase the production of this fruit. Millions of hectares of marginal land in arid parts of the world, where irrigation cannot be supplied, can come into their own as food producers when improved cultivars of cactus pear are available (Knight 1980).

Loquat

The loquat (Eriobotrya japonica, Rosaceae) is a small, evergreen tree native to central-eastern China, introduced into Japan in very early times. In Europe it was planted in the 18th century (Morton 1987). It is grown both as an ornamental and for its fruit. The total world crop was estimated as 150,000 t. It has been cultivated extensively in the Mediterranean basin (Spain, Algeria, Turkey, Israel, and Italy), Japan and China, to some extent in India and Brazil, and in a more limited fashion in Chile and the United States. Spain is the world's largest producer (40,000 t) and exporter of loquats, follow by Algeria (22,000 t), Japan (18,000 t), China, and Brazil (Lupescu et al. 1980; Morton 1987; Llacer et al. 1994). The tree grows best in a subtropical to warm-temperate climate. It does well on a variety of soils but does best on clay loams with good drainage.

The fruits, which are borne in large loose clusters, are commonly round, oval, or pyriform and in the best cultivars may reach a length of 7 cm. They vary in colour from pale yellow to deep orange and have a tough plumlike skin. The flesh is white to orange, firm or soft, juicy, and flavourful. From 1 to 4 smooth, brown seeds are commonly found in each fruit. The seeds comprise about 20% to 30% of the weight of the whole fruit (Insero et al. 1990). Loquats are consumed largely as fresh fruit, although small amounts are used in jams, jellies, syrups, and pies (Shaw 1980). Loquats are high in vitamin A and minerals. For the fresh market they should not be picked before full maturity; otherwise they are too acid. If properly handled they can be shipped to distant markets.

Generally, the loquat tree blooms in the autumn with fruits ripening in early spring (Apr.-May). They are normally pollinated by bees, but some cultivars such as 'Akko 13' from Israel and 'Golden Yellow' from India are not self-fertile, and others such as 'Advance' and 'Tanaka' are partially self-fertile. It has been observed that cross-pollination generally results in 10% to 17% increased production over self-pollination (Morton 1987). A list of important cultivars are: 'Advance', 'Algerie', 'Akko 13', 'Champagne', 'Magdall', 'Premier', 'Saint Michel', 'Tanaka', 'Thales' (syn. 'Gold Nugget'), etc. (Knight 1980; Morton 1987; Pathak and Gautam 1990).

The tree requires a minimal pruning, but some thinning of fruits may be required for optimum size. It is propagated mainly by budding or grafting onto seedling rootstocks; quince root can be used if a dwarf tree is desired. The tree comes into bearing in three to four years and the yield in the new orchards is very high (25 t/ha, Blumenfeld 1994). They are hand-harvested by clipping bunches of fruit and are sorted and graded by hand. The most important problems cause damage to the fruit rind, which downgrades fruit quality. In Europe two disorders damage the fruit rinds (Caballero 1993): sunburn ('purple spot'), a physiological disorder directly related to the calcium content in the fruit tissues, and the fungal black spot (Fusicladium eriobotryae). In California two diseases that sometimes create problems are fireblight (Erwinia amylovora) and loquat scab (Spilocea pyracanthae) (Ogawa and English 1991).

The overall production is steadily increasing. Several countries, such as Spain, Brazil, and India, are expected to increase commercial plantings of this crop. An interesting commercial aspect of loquat is that it ripens in early spring before other fruits (cherries, apricots, peaches, and plums) appear in the market. The most important problems (physiological and disease) result in a damaged fruit rind, which downgrades the fruit quality. Also, there is a need to reduce the labour costs of hand thinning and manual harvesting.

New Fruits for Arid Climates

In many countries around the world "developers" take the best agricultural lands for housing, and urbanization spreads rapidly. As a result, the agricultural industries are pushed into less productive lands in which the climate is not the most suitable for crops. A very good example is found in Israel. The best agricultural area is along the Mediterranean coast where the citrus industry flourished for almost a century and is the area where the famous 'Shamouti' (syn. Jaffa) orange was found and developed. Later, "Jaffa" became the trade name for the whole successful Israeli citrus industry. Today, most of this area is heavily populated and the agricultural industry is pushed into the Negev Desert where extreme temperatures exist, the soil is less fertile, and water, if available, is saline and/or very expensive. Labour is also less available and very expensive. The government by doing "more of the same" is encouraging the reestablishment of the citrus industry in the Negev and planting olives for oil (Negev of Growth Campaign).

Traditional crops in Israel appear to be at the end of their viable economic-life-cycle, but there is hope that new crops could establish a future profitable agricultural industry (Mizrahi and Nerd 1996). New crops could also ameliorate the increase in atmospheric CO_2 concentration (greenhouse effect) which increases temperature and drought, changing fertile lands into deserts (Mizrahi and Nerd 1996). The old traditional crops cannot cope with these expected changes. Over a decade ago Noel Vietmeyer (1986, 1990) raised the point that humans are "stupid" to neglect the huge treasure-house of bio-diversity available in the thousands of wild plants and neglected crops that might solve many of our present and future agricultural problems. The aim of our programme that started 16 years ago was to develop totally new fruit crops not known in the world markets, as new crops for desert areas (Nerd et al. 1990, 1993; Mizrahi and Nerd 1996, 1999). After many years of extensive R&D we had both failures and successes and we wish to share our experiences with others interested in new crops.

Fruit Trees Introduction

Since 1984 we have introduced and tested over 45 fruit tree species. Many of them are totally wild while others are neglected crops known only in their country of origin. All these species were tested in different ecozones around our Negev Desert, each differ from the other by average and extreme high and low temperatures and salinity of the irrigation water. In Israel, climatic and socio-economical considerations dictate that almost all crops are irrigated and in the Negev desert, most areas are irrigated with saline water. Salinity differs in the various ecozones not only by degree of salinity but also by ion composition. Whenever Na^{+} is more abounded than Ca^{++} (Rengel 1992) and Cl^{-} is more abounded than SO_4^{--}, the damage of salinity is more pronounced and more species will die even under 4dS/m2 (Nerd et al. 1990; Mizrahi and Nerd 1996).

Failures with Some African Species

As would be expected we have had many failures. Among our unsuccessful attempts is the yehib (Cordeauxia edulis Hemsl., Caesalpiniaceae), a wild shrub from the horn of Africa. This shrub produces a tasty edible nut even under extreme drought conditions when most plants will die. Unfortunately, we found yehib to be extremely sensitive to chilling temperatures, and plants have died at 4°C (Mizrahi and Nerd 1996). This unique bush deserves much more R&D efforts to save it from extinction (Miege and Miege 1979; National Research Council 1979).

Mongongo (Ricinodendron rautanenii Schniz, Euphobiaceae) is a nut-producing wild tree from the Kalahari Desert in southern Africa. It is highly nutritious and tasty nut but the yields are very low with no hope for the Israeli horticultural industry (Biesele et al. 1979; Fox and Norwood-Young 1982; Mizrahi and Nerd 1996). In all of the tested ecozones we only established fruiting mongongo trees in the Besor region with moderate temperatures and good quality irrigation water (Nerd et al. 1990).

A species which performed very well in most tested ecozones is the African plum (Harpephyllum caffrum Bernh. ex C. Krauss, Anacardiaceae). This wild tree of Southern Africa produces many small, tasty, aromatic, red fruits in bunches with different hues. However, there are problems that inhibit cultivation. First, the fruit is very small (few grams) and hence labour intensive and second the flesh is only 10% of the total weight. However it can be used for home gardens and city and park gardening with fruits that children and others will enjoy picking.

Cacti

Of the many plant families we have explored, the Cactaceae is the most important one for Israel, since water is the major limiting factor and becomes scarcer every day. In the future, the water authorities think that only recycled water will be used for agricultural production. The high water use efficiency of the cacti stems from their Crassulacean acid metabolism (CAM) pathway (Gibson and Nobel 1986; Nobel 1994; Mizrahi et al. 1997). Cacti have many uses as crops (Mizrahi et al. 1997). Our efforts have concentrated on fruit production of the unknown pitayas and not on the well-known prickly pear now called cactus pear [Opuntia ficus-indica (L.) Miller] that is already grown worldwide (Mizrahi et al. 1997, Mondragon Jacobo 1999, 2001).

Pitaya is a common name to many genera and species of cacti, all with elongated coloumnar stems (Ortiz 1999, 2001). Most of our work was recently reviewed (Mizrahi and Nerd 1999; Nerd et al. 2002) and here we would like to concentrate on updating the data and on the take-home-lesson and consequences from dealing with "real new crops." These crops are not well known and there is great confusion about their botanical identity. We shall describe briefly each one of the species that have made their way into the European markets.

Vine Cacti

Yellow pitaya [Selenicereus megalanthus (Schum.) Britton & Rose], is a vine cactus that needs a trellis system for support. In Israel net-houses are required to avoid photo-inhibition and bleaching of its stems (Mizrahi and Nerd 1999; Nerd et al. 2002). Colombia was the first country to sell S. megalanthus in the world market under the name "yellow pitaya." Until now, the incorrect names of Hylocereus triangularis and Hylocereus undatus have been used to describe this plant (Weiss et al. 1995). This plant can tolerate high temperatures more than the other vine cacti, yields spiny fruits where the spines abscise easily upon ripening and are unseen by consumers. The fruits are smaller than the other vine cacti fruits but the taste is superior, hence, the higher prices obtained in the markets in comparison with other vine cacti. Most of the plantations in

Colombia have been uprooted due to heavy infestation with fungi (Bibliowicz and Hernandez 1998). We found that desert environment is better than tropical, both from the phytosanitary point of view and the possibility to control and regulate plant production via irrigation, fertilization, and manipulation of shade regimes (Raveh et al. 1996; Mizrahi and Nerd 1999; Nerd et al. 2002).

The red pitaya [Hylocereus undatus (Haworth) Britton & Rose] is known in Latin America but the Asian name is dragon-fruit. There are red flesh clones but we do not know where they are produced. Some red flesh pitayas are produced in Nicaragua and are considered Hylocereus costaricensis (Weber) Britton & Rose. In Israel we cultivate Hylocereus polyrhizus Weber. Guatemalan researchers refer to it as Hylocereus undatus (Mizrahi and Nerd 1999). Pigments differ among the clones and species, and have not been fully explored. For example the Hylocereus sp. clone 10487 has red colour while the H. polyrhizus clones show glowing purple, a unique colour, which has been chemically identified as hylocerenin and iso-hylocerenin (Wybraniec et al. 2001). The major problem which exists in these plants in desert areas is their sensitivity both to low and high temperatures (Mizrahi and Nerd 1999). Since variability in these characteristics exist among genotypes, and since there is no genetic barrier among species and even genera, breeding may solve these problems (Lichtenzveig et al. 2000; Tel-Zur et al. 2001). Other important problems with these fruits which might be solved with breeding, is the bland taste and lack of distinct flavour. Sellers of these fruits in Europe claim that the dragon fruit and other red pitayas, which exist today in the market, are bought for their beauty and used mainly for decoration (A. Ronen unpubl. information). Our first hybrids released lately to our growers, exhibit much better taste than the original clones. However there is a lack of effort for breeding tolerance to extreme desert conditions, because of what we consider to be "establishment antagonism." We are convinced that in addition to marketing issues genetic and physiological R&D efforts are required to convert these exotics to mainstream commodities in the world markets.

Columnar Cacti

Among the various columnar cacti we believe that the most promising one is Cereus peruvianus Britton and Rose, which might be the same species as the Cereus jamacaru (Nerd et al. 1993; Mizrahi and Nerd 1999; Gutman et al. 2001). The plant is columnar with many branches. Columns may bear spines to various degrees and length, but the fruits are smooth and totally spineless.

The fruits are medium in size, and vary in skin colour from yellow to red with various hues. Flesh is white and aromatic with a delicate sour/sweet taste. Black seeds are embedded in the flesh, but are soft and edible reminiscent of kiwifruit seeds. Israel was the first country to sell this fruit both in the domestic market and in Europe. Efforts have been carried out in the US and Australia to introduce this unique fruit to their markets. This plant might have other industrial uses because of polysaccharides which may be extracted from its stems (Alvarez et al. 1992). Stem pruning is necessary to ensure efficient fruit harvest.

Our results in domestication of this fruit was recently reviewed (Mizrahi and Nerd 1999; Nerd et al. 2002) and here we would like to discuss difficulties in marketing which might hamper introduction. In Israel there are only a few producers. One producer is marketing the fruit at the proper stage of ripeness, with each package containing a leaflet describing the fruit, how to use it fresh, its other uses, storage and nutritional information.

However, others sell it in simple boxes with other fruits, at various stages of ripening, from unripe, to ripe and over-ripe, without explanatory leaflets, acts which damage the future of this newly developed fruit. In Europe columnar cactus fruit was accepted quite well but supermarket chains required 300 tonnes/year, while the quantities produced in Israel in 2001 were only 70 tonnes. Special efforts are required from the exporters to sell it in small quantities and if they are not determined to market this fruit it will not be sold. The Israeli establishment (as other establishments) is not interested in new crops. Despite these difficulties, we foresee a great future for this crop due to early and precocious yields, beautiful appearance, excellent delicate taste, long shelf-life to enable export, and, above all, minimum demand for irrigation and water-stress tolerance.

Other Fruit Trees Suitable for Desert Areas

Of the many fruit trees we have tested the following species are ready for the first trials to test their feasibility as new orchard crops under desert conditions. We have selected clones where we had information on tolerance to desert conditions, fruit yields, quality, and shelf life (Mizrahi and Nerd 1996). These fruits include white sapote (Casimiroa edulis Llave & Lex, Rutaceae) (Nerd et al 1992); black sapote (Diospyros digyna Jacq., Ebenaceae); desert apple ber or bor in India (Ziziphus mauritiana Lam., Rhamnaceae); marula from Southern Africa [Sclerocarya birrea (A. Rich) Hochst. sbsp. caffra (Sond.) Kokwaro, Anacardiaceae] (Weinert et al. 1990; Nerd and Mizrahi 2000); argan [Argania spinosa L. (Skeels), Sapotaceae], wild tree of Morocco, the best culinary oil is extracted from its seeds (Prendergast and Walker 1992; Nerd et al. 1994); and sapodilla [Manilkara zapota (L.) van Royen, Sapotaceae], from South East Asia, which is already a crop in many warm countries (Morton 1987). Some of these species (pitaya, white and black sapotes, and sapodilla) are also recommended for tropical zones such as Northern Queensland in Australia (Finocchiaro 2001). Unfortunately there is a lack of support to develop these and other new crops and farmers are afraid to take the risks. The daring farmers of Israel are already involved in production of various pitayas.

Another interesting fruit, which has not reached the R&D stage, is the monkey orange. It consists of three main species as follows: Strychnos cocculoides Backer; S. spinosa Lam.; and S. pungens Solereder, Loganiaceae, all native to Southern Africa (Wehmeyer 1966; Fox and Norwood-Young 1982; Taylor 1986). Strychnos pungens did not survive in any of our introduction orchards. Strychnos cocculoides which is considered the best of the three (in terms of eating quality), survived only in the Besor region (good quality water and moderate temperatures) and some trees started to bear fruits not of very high quality.

It is too early to judge its performance. So far the best of the three under our conditions is S. spinosa. It survived in three of our introduction orchards and performed very well in the Besor area. We have around 15 fruiting trees with high variability for growth, yields, fruit size, ripening season, and taste. Some of the seedlings bear astringent, bitter fruits, other bear very sour ones but two of the trees bear excellent tasty fruits. In organoleptic taste tests, people were requested to compare the monkey orange fruit with familiar fruits; the most common answers were, orange, banana, and apricot, and all possible combinations among them. The fruits emit a delicate aroma reminiscent of the spice clove. GC/MS analysis performed by Ephraim Lewinson of Newe Ya'ar, (ARO Israel) found eugenol, the essential oil found in clove (unpubl. results). Over 90% of the panel claimed that it was very tasty. Various products such as juices and dry fruit rolls are potential uses for this fruit. The fruit is large (400-1200 g), round, has a thick shell 4-7 mm, and contains 30%-45% juicy flesh with over 20% total soluble solids, and high acidity (over 200 µeq. H^+/gFW).

We are aware of the difficulties in introducing a new product into a market which is full of other temperate and tropical fruits. The first questions we are asked is-who needs these strange fruits? Who is going to buy them? For how much? What is the volume that the market can receive? All these are legitimate questions, which are discouraging for the establishment to deal with. On the other hand it is quite clear that apples and citrus fruits will struggle to withstand the changing economic and climatic conditions (Vietmeyer 1990). Also it is known that among the affluent class in the Western societies there is an increasing demand for exotic fruits and vegetables and some such as kiwifruit have become mainstream products (Anon. 2000).

The health profile of these new fruits is unknown and needs to be explored. For example the non-proteinecious amino acid, taurine was found in Opuntia ficus-indica, in spite of the common belief that the main source of taurine in our diet, comes from animals especially sea-food (Stitzing et al. 1999). Taurine is an important ingredient in the so called "energy drinks" with proven positive activity on human well-being (Seidl et al. 2000). In spite of all these arguments, exotic new fruit trees have received strong antagonism from all possible established organizations, such as the officials in the Ministry of Agriculture, Associations of Fruit Growers, and the main Export Company AGREXCO. We conclude with a plea from a report which speaks for itself.

"Every production system and every mature organization is equipped with antibodies against deviations from routine. These antibodies work overtime to kill all interest in the new product and are very likely to succeed. It is an innate property of every established industrial organization, which is why most new products are created and brought to market by young organizations that have no standard products to occupy them. Established organizations (Israeli agriculture among them) that want to get into this business of introducing new products have to build themselves managerial mechanisms designed to act as a countervailing weight against the forces of routine."

- Low price-elasticity for processed food products
- Need for distribution network and cold chain
- Backward-forward integration from farm to consumers
- Development of marketing channels
- Development of linkages between industry, government and institutions
- Taxation in line with other nations
- Streamlining of food laws

Deciduous Fruit

India produces all deciduous fruits including pome fruits (apple and pear) and stone fruits (peach, plum, apricot and cherry) in considerable quantity. These are mainly grown in the North-Western Indian States of Jammu and Kashmir (J&K), Himachal Pradesh (H.P.) and in Uttar Pradesh (U.P.) hills. The North-Eastern Hills region, comprising of the States of Arunachal Pradesh, Nagaland, Meghalaya, Manipur and Sikkim also grows some of the deciduous fruits on a limited scale. Due to introduction and adaptation of low chilling cultivars of crops like peach, plum and pear, they are also now being grown commercially in certain areas of the north Indian plains. Out of all the deciduous fruits, apple is the most important in terms of production and extent.

Apple was introduced into the country by the British in the Kullu Valley of the Himalayan State of H.P. as far back as 1865, while the coloured 'Delicious' cultivars of apple were introduced to Shimla hills of the same State in 1917. The apple cultivar 'Ambri', is considered to be indigenous to Kashmir and had been grown long before Western introductions. Pears and other deciduous fruits were domesticated successfully in the early part of the 20th century, although some of them were reported to occur under semi-wild conditions much earlier. Apricot was found growing in the drier pockets of northwestern Himalayas and two apricot varieties, locally known as 'Halman' and 'Rakchaikarpo' are reported to be indigenous to Leh-Ladakh areas of J&K State.

Sweet cherry was introduced from Europe before India's independence in 1947, while commercial cultivars of sour cherry have been brought mainly from USA in more recent years. The European and Japanese plum varieties are grown both in high and low hill areas. A plum variety 'Santa Rosa' reported to be a hybrid between Japanese and American species predominates (70-80%) plantations in the hills. Low chilling cultivars of peach and nectarine such as Flordasum, Flordared, and Sunred nectarine are successful introductions to the north-Indian plains. Some local selections of peach (Shan-e-Punjab, Sharbati), plum (Jamuni, Alubhokhara) and sand pear (Patharnakh) are also cultivated on a commercial scale in sub-tropical-marginal chilling areas of north India.

Deciduous Fruit Crop Cultivation

Crops Grown

India grows crops like apple (*Malus pumila Mill.*), pear (*Pyrus communis Berm.f.*), peach/nectarine (*Prunus persica (L). Bats. Ch.*), plum (*Prunus domestica L.*), apricot (*Prunus armeniaca L.*), sweet cherry (*Prunus avium L.*) and sour cherry (*Prunus cerasus L.*) on a commercial scale.

Cultivars

The promising cultivars of different temperate fruits in 3 major deciduous fruit growing States of India. Some relevant information on cultivars and cultivar selection is indicated below:

Apple: Over 700 accessions of apple, introduced from USA, Russia, U.K., Canada, Germany, Israel, Netherlands, Australia, Switzerland, Italy and Denmark have been tried and tested during the last 50 years. The delicious group of cultivars predominates the apple market. The areas covered under Delicious cultivars are: 83% of the area under apple in H.P., 45% in J&K and 30% in U.P. hills. In more recent times improved spur types and standard colour mutants with 20-50% higher yield potential are favoured. The important selections are:

- Spur types-Red spur, Starkrimson, Golden spur, Red Chief and Oregon spur.
- Colour mutants-Vance Delicious, Top Red, Skyline Supreme.
- Low chilling cultivars-Michal, Schlomit.
- Early cultivars-Benoni, Irish Peach, Early Shanburry, Fanny
- Juice making cultivars-Lord Lambourne, Granny Smith, Allington Pippin.
- Scab resistant cultivars-Co-Op-12, Florina, Firdous, Shirean.
- New Hybrids-Lal Ambri (Red Delicious X Ambri), Sunehari (Ambri X Golden Delicious), Amred (Red Delicious X Ambri), Chaubatia Anupam & Chaubatia Princess (Early Shanberry X Red Delicious) developed in India.

In H.P. monoculture of a few cultivars such as Royal Delicious, Red Delicious and Richared have started showing negative impact on the apple industry. Serious problems of apple scab disease and outbreak of premature leaf fall and infestation of red spider mite are causing great concern. U.P. Hills, particularly the Kumaon hills division, have the unique advantage of early harvest of apple, mainly due to cultivation of early maturing varieties like Early Shanburry, Fanny and Benoni. The early maturing varieties are harvested 2-3 weeks before the arrival of fresh apple from H.P. and J&K, and hence fetch very remunerative prices.

Pear: In pear for higher altitude conditions high chilling requirement varieties (like Bartlett) are mainly grown. In more recent years, red-colour strains of pear like Max Red Bartlett, Red Bartlett and Starking are replacing yellow coloured cultivars. In warmer sub-mountainous areas of H.P. and sub-tropical Punjab oriental pear cultivars like Baghugosha, Kieffer, China and sand pear Patharnakh are cultivated commercially both for table and processing purpose.

Apricot: Generally there are two types of apricot, namely, sweet kernel type and bitter kernel type. About 81 exotic accessions and 19 indigenous cultivars were collected and evaluated. The local types Halman and Rakhaikarpo have been popular, whereas exotic introductions namely, Nari, Kaisha, Shakarpa and New Castle are promising. These cultivars are recommended for dry cold areas. The USA variety Nugget is self-fruitful and bears sweet and attractive coloured fruits.

Peach: For colder conditions the peach cultivars July Elberta, Elberta, Peshwari, Quetta, Burbank and Stark Earliglo are well adopted. Low-chilling cultivars *viz.* Flordasum, Flordared, Shan-e-Punjab, Sharbati and Sunred (nectarine) have become popular in subtropical belts of U.P. and Punjab States.

Plum: A large number of cultivars (283) have been introduced into the country. European plums performed better in the hills, while Japanese plums are more adopted in sub-mountainous lower elevations. Leading cultivar in the hills is Santa Rosa. In the north-Indian plains small fruited cultivars like Titron, Kala Amritsari, Kelsey, and Alubukhara showed better performance. A good number of low-chilling Florida hybrids (Fla-1-2, Fla 73-4, Fla 85-2, 85-3, Fla 86-4) Sungold, Redgold etc., are under evaluation.

Cherry: Many cultivars of sweet cherry have been introduced from Europe, USSR and British Columbia. Promising exotic cultivars like Bigarrean Napoleon, Black Heart, Guigne Noir for J&K and cultivars like Black Tartarian Bing, Napoleon (white) Sam, Sue (White), Shella for H.P. have been identified. For warmer climate, cultivars like Summit, Sunburst, Lapins, Compacat and Stella have been found to be promising.

Promising Cultivars of Pome and Stone Fruits in Major Production Regions

Fruits	J&K	H.P.	U.P.
Apple	Benoni, Irish Peach, Cox's Orange Pippin, Ambri, White dotted Red, American Apirouge, Red Delicious, Golden Delicious	Tydeman's Early, Mollies Delicious, Starkrimson, Starking Delicious, Red delicious, Richared, Granny-Smith, Red Spur, Top Red, Red Chief, Oregon Spur, Golden Spur, Michal, Schlomit	Early Shanburry, Chaubattia Princess, Fanny Benoni, Red Delicious, Starking Delicious, Rymer, Buckingham
Pear	William, Kashmir Nakh, Vicar of Wakefield, Beuree D. Amanlis, Goshbagu, Beurre Hardy	China, Bartlett, Max Red Bartlett, Flemish Beauty, Devoe, Manning's Elizabeth, Winter Nellis	Thumb Pear, Doyene du Comice, Victoria, William's Bartlett, Beurre Hardy, Flemish Beauty
Apricot	Charmagz, Halman, Rachkaikarpo, Nari, Shakarpara	Kaisha, Nugget, Castle, Saffeida, Charmagz	Sharmagz Kaisha, Moorpark, Turkey, St. Ambrose
Plum	Satsuma, Santa Rosa, Burbank, Grand Duke, Titron	Kelsey, Santa Rosa, Titron, Satsuma, Mariposa	Jamuni, Kelsey, Santa Rosa, Titron
Peach	July Elberta, Elberta, Quetta, Flordasun, Shan-e-Punjab, Sharbati	Alton, July Elberta, JH Hale, Sharbati, Shan-e-Punjab, Burbank	Sharbati Safeida, Flordasun, Shan-e-Punjab
Sweet Cherry	Bigarreau Napoleon, Black heart	Black Tartarian, Napoleon, Sue Sam, Lambert Bing	Bedford Prolific, Black Heart, Governor's Wood

Rootstocks:

- Apple has been propagated mainly on seedling stocks of apple (*Malus pumila Mill*). Clonal rootstocks of M and MM series were introduced from East Malling Research Station of U.K. and M-9, M-26, M-4, M-7, MM-106 and MM-111 were identified as promising. Merton 779 was recommended as a commercial rootstock for apple for Kumaon hills of U.P. However, clonal rootstocks have not found favour among the apple growers. *In-vitro* protocols of M-9, MM-106 and MM-111 have been developed. Clonal rootstocks have been preferred only for high density orchards. Four rootstocks, namely, EMLA-111 for drought prone areas, EMLA-7 for slopy land and EMLA-106 for slopy and less clayey soil and EMLA-9 for high density planting on irrigated deep soils have been recommended. In pears, seedling stocks of *Pyrus pashia* (Kainth) is commercially used. Clonal rootstock Quince A with semi-vigorous growth has some promise. Newer clonal rootstocks like Old Home X Farmingdale series and BA 29C are under evaluation.

- For apricot, seedlings of wild apricot have been found to be good giving a smooth union and vigorous scion growth. Peach can also be tried as a rootstock, particularly for dry and light soils. Myrobolan plum is recommended for high moisture soil conditions.
- Peach is normally grafted on wild peach seedlings, plum (cv. Kabul Greengage) and even on apricot. Wild peach produced healthy and high-yielding plants. For controlling tree size, apricot rootstocks which impart dwarfing may be used.
- For plum, wild peach and wild apricot are commonly used as rootstocks. Among clonal rootstocks, Myrobolan B. was found to be most promising. Wild apricot seedlings and wild peach were also found to be suitable rootstocks for commercial cultivars like Santa Rosa, Mariposa and Beauty.
- Commercial plantations of sweet cherry were raised mainly on Mazzard and Mahaleb rootstocks. Seedlings are raised from late maturing cherry cultivars and stones of Waterloo or Misri cultivars are often used as rootstocks. Rooting of cherry suckers has been made possible with use of rooting hormones like IBA (6000 ppm) in combination with NAA (2000 ppm).

Area and Production

Deciduous fruits are mainly cultivated in North West Hills Region of India, comprising of States of Jammu & Kashmir (J&K), Himachal Pradesh (H.P.) and Uttar Pradesh (U.P.) and in the North Eastern Hills Region in the States of Arunachal Pradesh, Nagaland, Meghalaya and Manipur. Available data on area and production of apple and pear.

Area, Production and Yield of Apple and Pear in India

State	Apple (1995-96)			Pear (1992-93)		
	Area (ha)	Production (MT)	Yield (MT/ha)	Area (ha)	Production (MT)	Yield (MT/ha)
Arunachal Pradesh	5523	9730	1.76	938	2035	2.17
Himachal Pradesh	78296	276681	9.40	4653	8767	1.24
Jammu and Kashmir	78007	714834	11.0	14012	17941	1.28
Nagaland	64	109	1.70	460	1480	3.22
Uttar Pradesh	55200	210000	3.95	10550	27000	2.56
Total	217099	1211379	5.56	38536	175808	2.09

Apple: Although there has been 5-6 fold increase in apple production during the last 50 years, the productivity level is still very low (5.56 t/ha). Apple cultivation received greater attention by the growers. In H.P, area under apple increased from 3026 ha in 1960-61 to 78296 ha in 1995-96 with a corresponding increase in yield. J&K covers about 78007 ha under apple with a production of 714834 tons. In the U.P. hills (8 districts) apple occupies about 30 per cent of the area under fruits and contributes 46.9 percent of fruit production. The area covered under apple in U.P. hills is 55200 ha with production of 210000 tons of fruits. In the North-Eastern Hills Region, good quality apple is produced only in the rain-shadow belts of Arunachal Pradesh (5523 ha), and in Nagaland a very small area (64 ha) has been brought under apple cultivation.

About 99 percent of India's apple area falls under the North Western Hills region, covering 6 districts of J&K (Srinagar, Budgam, Pulwama, Anantanag, Baramullah, Kupwara), 6 districts of H.P. (Shimla, Kullu, Sirmour, Mandi, Chamba, Kinnaur) and 8 districts of U.P. (Almora, Nainital, Pithauragarh, Tehri, Pauri, Chamoli, Uttarkashi, Dehradun). In the Northeastern Hills region, good quality apple is grown in a small area in Tawang belt of Kameng district in Arunachal Pradesh. The Tawang area is basically a rainshadow belt and therefore, permits a longer period of sunshine and freedom from heavy rains, making it ideal for apple. Apple is also grown in Sikkim and Nagaland but the production is not a major success. Presently, a small quantity of apple produced in India is exported, mainly to Bangladesh and Sri Lanka.

Pear: Pear is grown under a wide variety of climatic regimes, ranging from cold dry temperate hilly conditions to warm humid subtropical conditions on the plains of northern India. Its cultivation is most extensive in J&K State (14012 ha), followed by U.P. (10550 ha). In the plains of Punjab, pear is grown in a specific area (7899 ha), whereas in H.P, it is widespread from the foothills to the highlands (600-2700 m) in areas experiencing 500-1500 chilling hours. Area, production and yield of apple and pear.

Others: There is no reliable data on area and production of other deciduous fruit crops. Stone fruits as a group occupy an area of 0.11 million ha, with 0.14 million tons production. In H.P about 28 thousand ha are covered under peach, plum, apricot, almond and cherry. In the Northwestern Himalayan region, peach holds greater promise because of its utilization for canning purposes. Peach is grown mainly in low and mid hilly areas (1000-2000 m above msl), except the low chilling cultivars belonging to the Florida group, which can be grown very well under sub-tropical conditions.

European plums require more chilling than Japanese plum. Japanese plums are generally grown in low and mid-hilly areas (1000-6000 m) conditions. The climatic requirement of apricot is almost the same as that of plum, while cherries require a colder climate (1000 chilling hours below 7° C). Cherry is commercially cultivated in J&K where the hailstorm problem is not encountered.

In the North-Eastern Hills (NEH) Region, these crops can be successfully grown in Arunachal Pradesh (Kameng, Siang, Tirap and Lohit districts), Meghalaya (Central plateau of Khasi & Jaintia hills), Manipur (Maram, Tadubi, Mou, Ukhrul) and Nagaland (Mokokchung, Wokhla, Thensang, Kohima, Phek districts). The productivity of all the stone fruits is low and estimated yield of peach is 0.50, 0.73, 2.11 and 2.23 t/ha in H.P, J&K, U.P, and NEH Regions respectively. The productivity of cherry in J&K is approximately 1.73 t/ha, while average yield of apricot is 0.42, 0.20 and 0.28 t/ha in H.P, J&K and U.P. hills respectively.

Production of Planting Material

Nurseries

There is a large number of government and private nurseries engaged in multiplication of planting material of deciduous fruit crops. In addition, the State Agricultural Universities and the Research Institutions multiply planting material of improved cultivars for sale and distribution to the farming communities. The Government of India had supported establishment of a large number of fruit nurseries, both in public and private sectors as planned activities. During the 8th Five Year Plan (1992-97) it is estimated that over 55 million nursery plants of different perennial crops, including temperate fruits, have been produced and distributed under the scheme. There was a target for the establishment of 85 big nurseries, 587 small nurseries, and 37 tissue culture units (20 by the private sector and 17 by the Government) of different fruit crops. Exact number of these nurseries engaged in production of pome and stone fruits is not known. However, the existing nurseries (more than 600), covering both public and private Institutions, are sufficient to meet the requirement of planting material of deciduous fruit crops.

Propagation and Rootstock Information

In all the pome and stone fruits vegetative propagation techniques of budding or grafting are followed for multiplication of planting material on standard rootstocks, raised both from seeds or through clonal methods. Cropwise details are as follows:

Apple: For raising rootstock seedlings, seeds of crab apple or commercial cultivars are stratified during December for 2-3 months at 2-5°C. One year old seedlings are used for budding/grafting.

Clonal rootstocks are raised through mound or stool layering. The mother plants are allowed to grow for one season and cut back to 3 cm from the ground level just before the growth begins. When new growth is about 10 cm, the shoots are covered with soil leaving the growing parts exposed. Rooted layers are cut off close to the ground level and planted in nursery beds for grafting/budding.

In H.P. hills, different types of budding/grafting are recommended:

- Chip budding in mid-June and mid-September
- Whip and Tongue and cleft grafting in February-March
- T-budding in May-June

Pear: Seedlings of Kainth (*Pyrus pashia Linn.*) or Shiara (*P. serotiana Rehd*) are stratified in moist sand for 35-45 days at 2-5°C. Tongue grafting in February-March or T-budding during June-July is recommended.

Others: Plum and apricot are generally grafted on wild apricot seedling rootstocks. Peach is also used as a rootstock.

Peach is budded/grafted on wild peach and peach-almond hybrid rootstocks. Hard wood cuttings (treated with 500 ppm IBA) can also be employed for raising rootstocks.

For apricot, peach, plum and cherry shield budding during July-August has been recommended for the hills of J&K.

Micro-propagation through tissue culture technique has not yet been commercialized for temperate fruits, although tissue culture protocols have been developed at research Institutes/Universities. Colt, a rootstock for cherry can be multiplied in large numbers by following the tissue culture technique. Shoot buds are cultured in MS medium containing 1-2 mg/liter benylaminopurine. Individual shoots are rooted on MS medium containing 1 mg/liter Indole Butyric Acid (IBA). The rooted plants are transferred to sterilized sand + soil (1:1) mixture and covered for 7-10 days for hardening.

Establishment of Orchards

Land Preparation and Planting

Fertile sandy-loam to clay loam soils with pH range from 5.5 to 7.5 and free from water logging conditions are suitable for establishing deciduous fruit orchards.

In flat lands/valleys, square, rectangular or triangular planting may be adopted. In hill slopes, planting on contours or terraces is recommended. In shallow slopy lands, small terraces (half-moon terrace), may be made to establish the plants and large scale disturbance of surface soil need to be avoided.

Planting of deciduous fruit plants is done during winter months from the end of December to mid March. Planting is done in the centre of pits (1x1x1m cube or circle), prepared a month before planting. While refilling the pit, 50g of aldrin powder (15%) is mixed with the soil.

After planting, the young plants are supported with stakes and basins are kept free of weeds. Mulching with dry grass or polyethylene is advisable. Irrigation is provided after planting.

Spacing

Spacing varies from species to species and depends also on the type of rootstocks used. For apple raised on seedlings of crab or other commercial varieties of apple, a planting distance of 6x6m or 7x7m accommodating 277-205 plants per ha is recommended for J&K hills. On clonal rootstocks like M4, M7, M26, MM106 a spacing of 4.5m x 4.5m (555 plants/ha) is suggested.

For spur type varieties and standard coloured mutants of apple, high density planting on dwarfing rootstocks like M9, M4, M7 and MM107 has been found to be feasible. Fruit yield of 30-35 tons/ha has been achieved in 12 year old orchards of colour mutants of apple on MM106 under a planting density of 2222 plants/ha (3m x 1.5m) in cooler hills of H.P.

Optimum Plant Spacing for Spur Types and Standard Colour Mutants of Apple

Type	Rootstock	Spacing (m×m)	No. of trees/ha
Spur Type	Seedling (Crab)	5×5	400
Spur Type	MM111, MM109	4×4	625
Standard Type	MM106, MM109	5×5	400
Spur Type	MM106, M7	3×3	1111
Standard Type	M9	2×2	2500

Source: Awasthi, R.P. and Chauhan, P.S., 1997. Apple and Pear. In: 50 years of Crop Science Research, ICAR Publication, New Delhi.

In pear, for trees on seedling rootstock under normal conditions, spacing of 5x5m is recommended, but on clonal rootstocks (Quince A) spacing can be reduced to 3x3m.

In peach, a spacing of 4.5 to 5m and for plum, apricot and cherry a spacing of 6m are followed.

Care and Management of Orchards

Training and Pruning

Modified central leader system of training has been recommended for both apple and pear on seedling rootstocks. The proportional heading back and thinning out system of pruning is adopted after the juvenile phase of plant growth. Spur pruning encourages vegetative growth and helps in new spur development in old plantations.

For high density planting on semi-dwarfing and dwarfing rootstocks spindle bush, dwarf pyramids and cordon system of training are suggested. Spindle bush on M7 and modified central leader on MM106 rootstocks are successful.

Peach is generally trained to open centre system, while in sweet cherry modified central leader system is followed. Pruning is done in such a way so that 20-50 cm new growth in young trees and 25-30 cm in older trees are secured every year. In sweet cherry about 10% of fruit bearing area should be removed annually, whereas in apricot slightly heavier pruning should be practised to keep the spur system renewed.

Manures and Fertilizers

Nutrient requirement varies from place to place. In Himachal Pradesh widespread deficiencies of N, P, K, Ca, Mn, Zn and B have been recorded. A fertilizer dose of 700:350:700 g of N, P and K for full grown bearing apple tree has been recommended.

It has been observed that VAM (Vesicular arbuscular mycorrhizea) fungi increases P uptake by apple roots. The following corrective measures for nutrient deficiencies have been suggested for apple.

Corrective Measures for Nutrient Deficiencies in Apple

Elements	Chemical & Dose	Time of Spray
N	Urea, 5.0%	Pre-petal fall
Ca	Ca Cl_2, 0.5%	30-45 days before harvest
Zn	$ZnSO_4$, 0.5%	After petal fall
Mn	Mn SO_4, 0.4%	After petal fall
B	H_3 BO_3, 0.1%	Before bloom and after bloom

In peach cv. Sharbati, 17.5 g N/tree/year and in apricot 70-100 g N/tree/year have been found to be optimum. For peach, N has been the main limiting element in H.P, while for plum, maximum fruit yield was recorded when leaf N, P and K contents were 2.89, 0.28 and 0.89 respectively.

Weeding and Mulching

In apple orchards, grass mulching (10 cm thick) with one application of post-emergence herbicide Glyphosate (0.8 kg/ha) was found to be effective. The shrubby weeds were, however, best controlled by 500 ppm Gramaxone + 100 ppm 2,4,5 -T application.

Moisture conservation was maximum under grass mulching and, therefore, recommended for apple in H.P. Mulching with oak leaf was effective in U.P.

Supplementary Irrigation

Trickle irrigation at 75 per cent of field capacity results in better tree growth and higher fruit yield in apple when raised on semi-dwarfing rootstocks. For apple, most critical period of water requirement is April to August and peak water requirement is just after fruit set. 114 cm of water during whole year through 19 irrigations have been recommended.

Drip irrigation saves water considerably. Field trials indicated a total irrigation requirement of 3840 liters water per tree under conventional system of irrigation; under drip system 1695 liters of water was enough. The application efficiency under drip system is about 2.27.

In Santa Rosa plum, irrigation at 50% field capacity gave better growth and economic returns.

Pest and Disease Control

Pests: In apple and pear about a dozen pests are causing serious damage to the crops. Most important ones are: San Jose Scale, Woolly apple aphid, Root borer, Blossom thrips, Coddling moth and European red mite. San Jose Scale can be effectively controlled with eco-friendly miscible spray oils at 2% concentration when applied during February-March. Woolly aphids can be controlled through soil application of Phorate or Carbofuran granules during May and October/November. For coddling moth, pheromone trapping was found to be effective, and certain bio-control agents have also been identified in apple orchards for controlling certain insect pests.

European red mite is becoming a very serious pest for apple in H.P. and J&K. The pest attack causes premature leaf fall. Late dormant sprays of miscible oil provide effective control of eggs. The mite can also be controlled by sprays of Dicotol (0.05%) followed by Malathion (0.05%).

In peach, leaf curl aphid can be controlled by pre-bloom sprays (at pink bud stage) of Dimethoate (0.03%) or Monocrotophos (0.04%).

Fruit fly in peach, apricot and plum can be controlled through foliar spray of baits consisting of Malathion (0.1%) + 1% sugar.Stem and shoot borers causing damage to peach, plum, apricot, almond and cherry can be controlled by inserting 0.5g of PDCB (paradichlorobenzene) into the holes and plugging them with mud.

Diseases: Apple scab caused by *Venturia inaequalis* is a serious disease causing maximum economic loss. A sound forecasting and early warning system has been developed for prediction of scab attack. Also, a judicious fungicide spray schedule has been devised. Under high disease pressure, systematic fungicides performed better, while under low disease pressure Ergosterol biosynthesis inhibiting (EBI) fungicides were as good as protectants. Ascosporic inoculum produced by over-wintered apple leaves could be substantially reduced by giving post-harvest applications of Bavistin (0.1%) and EBI chemicals Penconazole (0.5%) and Flusilazole (0.01%) as pre-harvest fungicidal sprays control scab during storage.

Powdery mildew disease can be kept under check by pruning and spraying wettable sulphur (0.2-0.3%) or Karathane (0.05%) during dormancy, bud swell and petal fall stages.

Collar rot and white root rot diseases in apple occur mainly in poorly drained soils. Proper drainage of orchards and soil drenching with 0.1% Carbendazim for white root rot and with 0.3% Mancozeb or 0.3% Ridomil MZ for collar rot are effective. Virus and virus-like diseases such as mosaic, chlorotic leaf spot, rubbery wood and others have been reported. Virus cleaning through tissue culture and supply of virus free bud wood material are being pursued to contain further spread of the viral diseases in apple.

Intercropping

As apple is grown in hill slopes, seasonal intercropping accompanied by working the soil is not encouraged in order to prevent soil erosion. Where slope of the land is more than 10 per cent, various grasses and legumes are grown as permanent covers. The tree basins are kept clean and sod culture in the orchards is practised in the hills of H.P. The grasses are regularly cut when 7-10 cm high and the residue is left in the orchards for decomposition.

The cultivation of legumes and grasses as intercrop is becoming popular in orchards. The important legumes are red clover (*Trifolium pratense*), white clover (*T. repens*) and lucern (*Medicago sativa*). Among the grasses, orchard grass (*Dactylis glomerata*), fescues (*Festuca arundiceae*) and Timothy (*Phleum pratense*) are common. Orchard grass is grown where moisture is high, while fescues is suitable for dry areas or southern aspects. In apple, sometimes filler trees of peaches or dwarf apple plants are grown.

At the young age of orchard plantations, vegetables like cole crops, potato and tomato are grown to supplement the income. However, this practice is abandoned when the orchard starts bearing after 5-7 years to avoid competition. In Shimla hills and Kullu Valley areas of H.P. cultivation of gladioli for flower and bulb production as intercrop is carried out in the apple orchards since it can tolerate shading.

Harvesting and Yield

Apple: In apple, the number of days taken from full bloom to harvest are 132-134 days for Starking Delicious, 138-140 days for Red Delicious and 147-148 days for Golden Delicious varieties in the high hills (above 2000 m) of H.P. In J&K State, the variety Maharaji requires 160 ± 3 days after full bloom for proper harvest maturity. Apple cultivar Granny Smith takes 180 ± 5 days for maturity.

Maturity Indices for Commercial Cultivars of Apple

Cultivar	Days from full bloom	Firmness (kg)	TSS (%)
Royal Delicious	125 ± 5	8.2 ± 0.40	13.0 - 15.0
Red Gold	122 ± 3	8.3 ± 0.20	12.0 - 13.5
Red Delicious	134 ± 5	8.4 ± 0.40	10.0 - 14.0
McIntosh	135 ± 4	6.8 ± 0.25	11.5 - 13.5
Golden Delicious	148 ± 6	8.4 ± 0.40	12.0 - 14.5
Granny Smith	180 ± 5	8.7 ± 0.30	11.5 - 13.0

Fruit Drop Control: Pre-harvest fruit drop in apple could be effectively checked with the application of 10 ppm NAA 20-25 days before harvest. Pre-harvest fruit drop is quite serious in early ripening cultivars like Tydeman's Early, Red Gold and Pippins, where loss of 40-60% crop load is often experienced. Mid-season Delicious group cultivars also experience 15-20% loss.

Colour Development and Enhancement of Ripening: In apple, fruit colour development in warmer and lower (below 1800 m) elevations is generally poor. Application of Ethephon (2-chloroethyl phosphonic acid) about 10 days before harvest improves red coloration. Induction of early maturity by 7-10 days in mid hills situations could be achieved through an application of 500 ppm of ethephon + 10 ppm of NAA 3 weeks before harvesting. In H.P, two pre-harvest sprays of 0.5% $CaCl_2$, followed by a 0.03% surfactant like Tween 20 for 1-2 minutes have been recommended for better shelf-life of apple.

Fruit Thinning: In apple, for obtaining better grade fruits of optimum size a post-bloom spray of 100-200 ppm ethephon 2-3 weeks after full bloom has been recommended both in H.P. and J&K.

Pear: Harvesting time of pear is determined by slight change in fruit colour and easy separation of fruits. Fruits should not be allowed to ripen to soft condition on the tree and should be harvested when they are still hard. Depending on cultivars fruit may be ready for harvest from 70 days after full bloom (*viz.* cv. China pear) to 135 days after full bloom (*viz.* cv. Barlett). The Bartlett pear normally takes 122 ± 3 days in valley (Kuala) areas of H.P.

Cherry: Sweet cherry should be harvested when fruit stalk separates easily from the spur and taste develops typical to the cultivar. From full bloom it takes 45 days in cultivars like Guigne Noir and Gross Lucenta, whereas in cultivars like Bigarrean Napoleon the number of days varies from 65-75 days.

Peach: For harvesting peach at the right time, the proper colour development in fruits and pit browning are considered as reliable guides. In yellow fleshed cultivars, deep orange colour development on fruits is associated with proper maturity. At least 5% of the pit area also should be brown if the fruit is to develop good flavour at ripening. In the peach variety July Elberta, it takes about 90-95 days after full bloom, whereas on cv. Elberta it normally takes 100-105 days. In low chilling cultivars like Flordasun, fruits should be harvested at 50% colour development stage.

Plum: In free-stone cultivars of plum, pit browning up to 5% of the pit area is a good index of maturity. For proper fruit maturity, early cultivars like Sharp Early and Formosa take 95 to 105 days, mid season variety Santa Rosa requires 115-120 days and late cultivars like Grand Duke take 130-135 days after full bloom.

Apricot: Apricot fruits of a single tree ripen for a period of about 3 weeks. The fruits develop detectable flavour when allowed to ripen on the tree. The early cultivar Charmagz matures within 75 days from full bloom, while the late cultivar Quetta requires 125-130 days for proper harvest maturity.

Yields

There is a wide range of variation in fruit yield in different States and in different cultivars. The average yield of apple and pear have been indicated earlier. Although average productivity of apple in H.P. has been shown as 9.4 t/ha, some well managed orchards of Delicious apple in Kotgarh area of the same State yield about 50 t/ha. Similarly, yield record of 35 t/ha has been achieved in 12 year old plantations of coloured mutant apple cultivars on MM 106 rootstocks under high density planting system (2222 plants/ha) in a research station in Shimla hills of H.P.

Average Yield of Different Stone Fruits by State

State	Fruit Yield, T/ha			
	Peach	Plum	Apricot	Cherry
H.P.	0.50	1.31	0.42	-
J&K	0.73	0.74	0.20	1.73
U.P.	2.11	2.50	0.28	-
NEH Region	2.50	1.18	-	-

Marketing: The biggest wholesale market for apple is the Fruit and Vegetable market at Azadpur, in Delhi. About 70 per cent of the total trade of apple is distributed through this market. The seasons of market arrivals from North Western States are as follows:

H.P.-July to October with a peak in August-September

J&K-August to November, with a peak in September-October

U.P.-June to October, with a peak in July-mid September.

There are a number of marketing channels, of which the predominant are:

(a) Farmer-Pre-harvest contractor-Commission agent-Wholesaler-Retailer-Consumer.

(b) Farmer-Forwarding agent-Commission Agent-Wholesaler-Retailer-Consumer.

(c) Farmer-Commission agent-Wholesaler-Retailer-Consumer.

The sale through pre-harvest contractors is the most important system of marketing. Normally, the small orchardists sell their crop at flowering stage to contractors who organize plant protection practices, picking and packaging of fruits.

The medium and large orchardists prefer to market their produce through channels (b) and (c). In H.P., about 65% of total apple produced is marketed through these two channels. Some Growers' Cooperatives and Government controlled marketing Corporations (like HPMC in H.P.) also get involved in the apple market. For example, in H.P the volume of trading handled by Growers' Cooperatives is about 3.5 to 4.0 per cent, while another 2-2.5 per cent of the total produce is marketed by HPMC.

The National Horticulture Board (NHB) of the Government of India regularly publishes and announces the wholesale price and market arrival figures of apple in different terminal markets for the benefit of the growers. Some quantities of apple are placed in cold storage facilities.

The Himachal Pradesh Horticultural Produce Marketing and Processing Corporation Ltd. (HPMC) has set up modern fruit packing houses in different parts of the State. Each packing house has a capacity to grade and pack 55,000 to 275,000 cartons of fruits. Cold storage facilities at the production centres and terminal markets at Delhi, Bombay (Mumbai), Madras (Chennai) and Calcutta have been created. Similarly, the HPMC has undertaken large scale processing of temperate fruits like apples, peaches and others.

In H.P, grading and packing of fruits have been organized reasonably well. In apple, 7 size grades have been prescribed and dimensions of packing boxes, size of wrapping papers and number of layers of fruits for each grade have been standardized. In apricot, plum and peach, three size grades have been prescribed. Presently CFB (Corrugated Fibre Board) telescopic tray pack cartons for apple and smaller size universal CFB cartons for stone fruits are in use. These are available with HPMC at their various packing stations. Plastic crates (both collapsible and non-collapsible) are also used for cold storage and for processing units.

Picking and packaging of fruits are done in different ways at different sites. In U.P hills, trees are generally strip picked into small wicker baskets or small gunny bags which are emptied into larger baskets for carrying to packing stations. Both U.P.

and J&K have developed packaging units for different temperate fruits.

India presently exports a small quantity of apple (5.95% of total fresh fruit export), mostly to Bangladesh and Sri Lanka. Apple cultivars like Red and Royal Delicious, Ambri and other new coloured cultivars are suitable for the export market. According to the Trade Year statistics of 1993, India exported only 7000 tons of apple during that year.

Processing

The HPMC in Himachal Pradesh owns two fruit processing plants with a combined capacity of 30,000 tons annually. The plants are equipped with modern machinery for six-fold juice concentration and aroma recovery. The plants have also facilities for aseptic bulk packaging and pasteurization. The HPMC manufactures mainly primary products like fruit pulps, single strength juices and juice concentrates. It covers fruits like apple, pear, peach apricot and plum. The pomace of the fruits after juice extraction is used for cattle feed, extraction of pectin etc.

In J&K, low grade fruits are processed for various products like apple juice, jam, jelly, fruit bars, rings, vinegar etc. from pome and stone fruits. In U.P. hills also, various types of processed products are prepared from temperate fruits. The units are mainly small and medium in size and not to the scale of HPMC. On the research side, important cultivars of apple were screened for their suitability for preparation of juice, jam, dehydrated rings, chops, cubes and preserves. Golden Delicious cultivar was found to be the multi-use cultivar to meet the FPO quality requirements.

A new process for canning of apples called osmo-canning has been developed. Similarly, process for preparation of cider, wine and vermouth have been developed. Utilization of apple pomace for production of pectin and natural colours have been attempted. Although CA storage concept has not yet become popularized, fruits placed in CA store at Shalimar in J&K have shown that this system of storage has great promise for temperate fruits.

Potential for Deciduous Fruits

Area Expansion: Both North-West and North-Eastern regions of India offer large areas ideally suitable for cultivation of pome and stone fruits. In the North-West India, stone fruits like peach, plum and apricot come up well at elevations between 900 to 1500m with an annual rainfall of 90-100cm. Apple, cherry and pear are commercially successful at elevations between 1500-2700 m above msl. The cold arid regions between 1550-3650m with annual rainfall of 25-40 cm are again highly suitable for dry nuts and drying type of apricots. There are vast tracts of land still available for further expansion of these crops.

In the North-Eastern Hills, excepting the rain shadow belts of Arunachal Pradesh and high altitude Lachung area of Sikkim, apple may not be successful commercially. Other crops like pear, peach, plum and apricot offer good scope for further expansion.

Widening the Cultivar Base: The low productivity and poor quality of apple are often linked with monoculture of a few old cultivars and their degeneration over the years. The U.P hills, particularly Kumaon division, has the unique advantage of harvesting apple fruits for early market. Similarly, the rainshadow belts in North-Eastern Hills can offer good quality apple for the eastern Indian markets, thus reducing the cost of long distance transportation from North-Western Hills. Part of the markets of Bangladesh can be captured by the fruits of the North-East.

There is good scope for introduction of new promising cultivars, replacing the Delicious group. Similarly, use of clonal rootstocks of Malling and Malling-Merton series and even their indexed material 'EMLA' selections will greatly change the productivity and quality of fruits. High density planting with spur type cultivars offer good scope. The identification of low chilling peach, plum and pear cultivars offer good possibilities for their cultivation in the low hills and in sub-tropical plains. Some of the new hybrids, including scab resistant apple cultivars need verification trials on a commercial scale.

Management Practices: Scientific water management and practising proper training and pruning of trees including introduction of renewal pruning techniques, will make significant impact on increased production even in the existing orchards. Drip irrigation, *in-situ* water harvesting and correcting macro and micro-nutrient deficiencies will go a long way in bringing notable improvement in productivity as well as fruit quality. By adopting IPM strategy and organic farming practices, selected export markets can be targeted well.

Processed Products: There is immense scope for increasing various processed products of pome and stone fruits, for which technologies are available. The existing capacity of the most organized processing unit of HPMC is only marginal as compared to the volume of fruit available for processing. The HPMC utilized only 1 to 1.5 percent of total cull fruit available. The present combined capacity of two units of HPMC is 30,000 tons annually, which can be easily raised to 50 to 75 thousand tons.

The expansion of grading and packing stations, their further modernization with mechanical grading equipment, use of CFB boxes and more number of pre-cooling and cold storage units will improve the marketing system and enhance marketability of the produce. Apple has been identified as one of 6 most promising fruits for fresh fruit exports.

Constraints in Deciduous Fruit Production

Large number of old orchards (more than 30 years old) are showing decline in terms of growth and fruit yield. Such old trees do not produce adequate extension growth. Large scale replanting is therefore needed. Delicious group of cultivars constitute the major share (about 83% in H.P.) of apple production in the country. These cultivars are self unfruitful and need cross pollination to ensure good fruit set. Interplanting pollinizer cultivars (Golden Delicious, Jonathan, Red Gold, Lord Lambourne etc.) in the proportion of 25 to 33 percent is necessary for good fruit set, and choice of wrong pollinizers and their inadequacy in number often result to low productivity.

In many countries, Delicious group has been replaced or is in the process of replacement with more promising cultivars.

The need for injecting new blood into the apple industry through spread of new cultivars (spur types, colour mutants, strains of Gala, Red Fuji; scab resistant cultivars, bud sport selections of Royal Delicious, and some of the promising hybrids) is urgently felt. Some of the spur type and coloured mutants are already popular with farmers and high density planting has also caught the imagination of developmental departments and agencies both in H.P. and J&K. The research system has already identified Early, Mid and Late cultivars for different agro-climatic regions.

The low chilling cultivars of stone fruits have also covered large tracts of the sub-tropical plains of Punjab, U.P and H.P. For the hills, promising cultivars identified need further spread.

Generally, apple is grown in marginal land and fertilizers are not applied according to the requirements of the trees. The water and fertilizer use efficiency is generally poor. Also, spring frost and hailstorms are adverse weather parameters leading to low productivity. Research results have shown that through proper orchard management practices (soil and water conservation and fertilizer application) the fruit yield can be doubled in the existing orchards. The adoption of improved production technology developed by the research system can bring visible and perceptible changes in the temperate fruit industry in India. Technologies like use of clonal rootstocks, introduction of renewal pruning techniques and micro nutrient applications have not been transferred and adopted at a satisfactory level.

Apple scab disease has been the major plant protection problem in apple in J&K and H.P, whilst U.P hills are comparatively free from the disease. Apple scab forecasting system developed and the chemical control schedule prescribed have helped in reducing loss due to apple scab to a considerable extent. Apple growers are adopting the prescribed schedule of chemical sprays to control the disease. For checking entries of diseased material in the free areas of U.P. and North-Eastern Hills, strict quarantine and selection of elite disease-free mother plants are very essential. Often it is not followed strictly. Some of the virus diseases have also been reported in apple for which biological and serological indexing/detection procedures have been developed. Limited quantity of virus-free budwood is also being supplied. Extreme care is now required to be taken to multiply quality planting material (in apple alone approximately 2 million plants/year) for establishing new plantations.

Most of the orchardists still sell their crop at flowering to contractors as there is no well organized marketing system. Transportation in the hills itself is problematic. Post-harvest management problems originating from poor harvesting (strip picking) and improper packing system (non CFB boxes) and lack of proper pre-cooling and cold storage facilities result in huge (25-30%) loss of fruits. Capacity of the processing sector is also inadequate. Product diversification, value addition and market infrastructure development would require very substantial investment. The existing processing units are quite old and they require modernization for which substantial investment is required. CA storage trials have shown good promise. Its extension in larger growing areas is needed. Technology for storage of apple is now known, as a result of which apple is now available throughout the year.

Government Policies and Plans

Research on pome and stone fruits is conducted mainly by three State Agricultural Universities, namely: (a) Sher-e-Kashmir University for Agriculture Science & Technology, J&K; (b) Y.S. Parmar University for Horticulture and Forestry, H.P.; and (c) G.B. Pant University for Agriculture & Technology, U.P. A good number of research stations of these Universities located in major pome and stone fruit growing belts are engaged in temperate fruit research. The Indian Council of Agricultural Research (ICAR) through All India Coordinated Research Projects on Fruits and Post-harvest Technology and another Network project on Apple Scab disease has further strengthened the research activities in deciduous fruits. A few long established temperate fruit research stations namely at Shalimar Bagh in J&K, at Mashoobra in H.P. and at Chaubatia in U.P. hills have made commendable progress in temperate fruits. Both State Governments and the ICAR provide financial support to strengthen fruit research in these Universities. A large number of ad-hoc research projects on pome and stone fruits funded by ICAR are also generating good information on these crops.

During the 8th Government Developmental Plan (1992-1997) the ICAR established a Central Institute for Temperate Horticulture (CITH) with its headquarters at Srinagar in the J&K State, with a regional station at Mukteshwar in the U.P. Hills. Both these research stations will work exclusively on temperate fruit crops. This new Institute will receive major support during the 9th Plan period (1997-2001). In the North-Eastern Hills region, the ICAR Research Complex for NEH Region with its headquarters in Meghalaya State and regional units in each of the 5 other States are engaged in fruit research.

On the developmental side the State governments are engaged in nursery production of quality planting material. For example, in H.P. alone currently there are 600 nurseries in private and public sector producing and distributing more than 0.8 million plants of apple alone, every year. There are ambitious programmes in all the States to further expand/ replant with new improved cultivars. Apple scab disease control and post-harvest processing sectors are getting focused attention in Government developmental plans. The Directorate of Marketing and Inspection of the Government of India has framed grade standards for apple, plum and William pear. The organizations like National Horticultural Board (NHB), National Cooperative Development Corporation (NCDC), Agricultural and Processed Food Products Export Development Authority (APEDA) etc., are providing incentives to traders and exporters to improve their infrastructural facilities like grading and packaging centres, refrigerated transport, setting up of pre-cooling, cold storage, auction platforms etc. The NCDC is undertaking procurement and marketing of apple on a limited scale. The NHB has set up a market information service for the benefit of growers.

Conclusion

Deciduous fruits, covering pome and stone fruits contribute significantly to the horticulture economy of India. Apple production dominates the scene and systematic cultivation and marketing of apple can change the rural economy in the hills of North-Western India. New vision and concerted efforts are required for change in variety mix, supply of quality planting material from elite clones on indexed clonal rootstocks. High density planting, water management including micro-irrigation, integrated plant nutrient management and IPM strategy for plant protection are some of the areas which need greater R&D focus. Adoption of post-harvest management practices and infrastructure development for grading, packaging, pre-cooling and storage of the produce needs focused developmental attention. Value addition and export promotion, particularly of apple are drawing due attention of the developmental agencies in India.

Ginger, Fat and Fibrinolysis

Ginger (Zingiber officinale) is a popular food spice, and is used as a medicine to treat various ailments from time immemorial in different parts of the world. It occupies an important place in Ayurvedic and Graeco-Arabic systems of medicine where it is commonly used as carminative, digestive and to treat chronic rheumatism and gout. It is customary in India to add ginger in the diet rich in fat. It is reported to contain anti-histaminic and antioxidant factors.

Earlier, it has been reported that dietary condiments and spices such as garlic, onion and asafoetida neutralize fat induced alteration in blood lipids, platelet functions and fibrinolysis. Ginger has also been evaluated regarding its effect on blood lipids and platelet aggregation both in healthy individuals and patients of coronary artery disease, However, its effect on fibrinolytic activity has not been studied. The present study therefore, was envisaged to observe the effect of ginger on fibrinolytic activity (FA) modified by high fat diet in healthy individuals.

Material and Methods

The study was conducted on 30 healthy adult males, between the age of 30 to 50 years who were not consuming tobacco in any form. After a written consert, fasting blood samples were collect and they were immediately administered 50 gm of butter with four slices of slices of bread. Blood samples were again collected after 4 hours. During the subsequent week, a similar procedure was adopted on the same subjects on 2 days, but one of the following was administered randomly in addition to the butter: (a) Ginger 5 gm in gelatin capsules or (b) placebo in matched gelatin capsules. Blood samples were again collected. Thus all 30 patients received 50 gm of butter t 5 gm of ginger on one occasion and 50 gm butter + placebo on other occasion. Therefore, all served their own control.

All the blood sample were analysed for fibrinolytic activity which was determined by employing a method described by Buckell & Elliot. The method is based on the following principle: the euglobin fraction of plasma contains plasminogen activator, plasminogen and fibrinogen. Normally occurring inhibitors of conversion of plasminogen to plasmin are not present in this plasma fraction. The euglobulib fraction is clotted with thrombin and the time taken for clot lysis is estimated and expressed in units by multiplying the reciprocal of the lysis time in minutes by 10,000. The results were statistically analysed with student's 't' test for paired data.

Preparation of Ginger Capsules: Dry ginger rhizomes were powdered in the mortar and pastle. The dry powder was filled in gelatin capsules. Each capsule contained 625 mg of dry ginger powder. Eight capsules were administered with butter. Placebo capsules contained same amount of lactose powder.

Results

Fibrinolytic activity decreased by 18.8 per cent after administration of fatty meal. Addition of ginger along with the fat not only prevented the fall in fibrinolytic activity but actually increased it by 6.7 per cent. As compared with fatty meal, ginger has actually increased increased fibrinolysis by 31.5 per cent. The placebo has not prevented the fat induced decrease in fibrinolytic activity.

Discussion

Administration of 50 gm of butter has lead to a significant fall in FA ($P < 0.001$). Addition of ginger to a fatty meal in the dose of 5 gm not only prevented the fat induced decrease in

FA. But kept the fibrinolysis above the normal level. The change is significant ($P < 0.001$) as compared to control experiment in which fatty meal was administered alone to the same subjects. Preliminary trials have lesser amount of ginger would give inconsistant results therefore 5 gm dose has been used throughout this study. No side effects were observed even in this dose.

Dietary fat administration leading to significant fall in fibrinolytic has been observed by many workers. A rich fatty meal if consumed frequently for long time even in apparently healthy individuals may be harmful because it may lead to poor tendency to clot lysis. Moreover, this defective fibrinolysis may be an important factor in the genesis of coronary artery disease as has been stressed by several workers. Therefore, addition of ginger may prove to be a convenient and safe dietary measure for every day use in persons predisposed to such situations.

Chemically ginger contains several classes of compounds. The chemical composition of dried ginger is as follows: starch 40-60%, proteins 10%, fats 10%, fibres 5%, inorganic material 6%, residual moisture 10% and essential oil (oleorasin) 1-4 per cent. The essential oil of ginger contains various terpins and sesquiterpenes. The predominant sesquiterpene hydrocarbon is zingiberene. The characteristic pungent odour is due to its oleoresin content which is an oily liquid containing oxymethyl phenols like shogaol, zingerone and gingerol etc. In all more than 200 different volatile substances have been characterised in the essential oil fraction; out of all which compound(s) is exactly responsible for frbrinolysis is not known.

Ginger has been a popular food spice possessing many medicinal properties. Recently it has drawn a lot of scientific attention. It has been reported to have components (gingerols, shogaols) which are potent inhibitor of platelet aggregation and inhibit cyclooxygenase and 5-lypoxygenase in vitro. It has also shown amelioration in symptoms (pain and inflammation) in patients with musculokeskeletal disorders and migraine However, it does not affect blood lipids, blood sugar and fibrinogen in patients with coronary artry disease. The present study, therefore, brings out a new property of ginger, not reported earlier, whereby it neutralizes the altered fibrinolytic state induced by fatty meal in healthy individuals.

We have earlier reported that ginger inhibits platelet aggregation in healthy individuals and patients with coronary artery disease. The effect on fibrinolytic activity further adds to its therapeutic value. Moreover it also confirms the rationale of the age old Indian tradition of adding various condiments to a diet rich in fat in order to neutralize the deleterious effects of a fatty meal. Ginger may therefore prove useful in situations which are associated with altered state of fibrinolysis.

Summary

Administration of 50 gm of fat to 30 healthy adult volunteers decreased fibrinolytic activity from a mean of 64.20 ± 531 to 52.10 ± 3.20 units ($P < 0.001$). Supplementation of 5 gm of ginger powder with fatty meal not only prevented the fall in fibrinolytic activity but actually increased it significantly ($P < 0.001$). This fibrinolytic enhancing property is a further addition to the therapeutic potential of ginger.

Revitalisation of the Indigenous Flora

The indigenous vegetation of the Coromandel Coast bio-region is known as Tropical Dry Evergreen Forest. This forest type is found only in South India and Sri Lanka and provides a rare biological richness due to its very high species density, but it is now close to extinction.

There are only three small remaining areas (approximately 4000 acres) of this forest type in the bio-region and although they receive protection from the State Forest Department there is no official policy to revitalise them. There are more than 7000 species of plants used for medicinal purposes by the people of India. This medical heritage represents one of the longest unbroken traditions in human civilisation.

There are two main streams of medicinal knowledge in India; the local folk medicine located in the villages, and the codified knowledge systems of Ayurveda, Sidha, Unani and Tebatten. Of these two systems, local medicine is perhaps the most practical and accessible for basic health care needs in the villages. This particular bio-region is home to at least 400 species of medicinal plants that have been used for centuries by traditional practitioners of herbal medicine. However, the few remaining areas of TDEF are under constant threat and genetic resources are being rapidly depleted from, for example, clearing by forest departments to plant monoculture exotic fuelwood species or from over exploitation by the pharmaceutical industry which collects 95% of its plant materials from wild sources such as these.

A few local, plant doctors still know the uses of many of the medicinal plants of these areas and in some cases the knowledge has been preserved on palm leaf manuscripts, some written eight generations ago. However much of the knowledge is being lost in a negative cycle of cause and effect stemming from the degradation of the TDEF = lack of raw drugs = lack of practice = depletion of skills = cessation of networking between practitioners = loss of self esteem = lack of young people taking interest in learning the skills.

Strategies for Revitalization

Within the bio-region is the international township of Auroville which was established in 1968. Land restoration both in and around Auroville has been an ongoing process since this time, with the result that "new" forests have been established and over 2 million trees planted. Pitchandikulam is a community within Auroville which has been heavily involved in this "Greenwork", particularly with regard to medicinal plants. Pitchandikulam is now, after 25 years, a diverse eco-system containing more than 400 species of plants covering an area of approximately 50 acres. Windmills, Biogas and Solar Power are used for the communities energy needs and the gardens, orchards, woodlots and forest sanctuaries are often used as training sites for many aspects of eco-restoration work.

In 1992, The Foundation for Revitalisation of Local Health Traditions (FRLHT) incorporated Pitchandikulam as one of its Medicinal Plant Conservation Parks. FRLHT is a non-governmental organisation focusing on re-building India's vast traditional medical knowledge. In an effort to focus attention on India's ethno-medical heritage, one of the programmes that FRLHT has embarked on is that of conservation as well as sustainable use of medicinal plants, especially in the area of primary health care. The foundation's thrust areas include establishing a computerised information network on medicinal plants and natural products, promoting the utilisation of traditional medicine in rural and urban areas and establishing cooperative links with other traditional medicinal systems of the world.

In 1993, Pitchandikulam Bio-resource Centre was established as a focal point for indigenous forest regenerative work in Auroville with special emphasis on the medicinal plant aspect and interaction with the local community, particularly traditional plant doctors, school children and other interested organisations and individuals.

An important aspect of this work is training of community health workers, village development workers and school children. The centre displays raw drugs, seeds, examples of local traditional technologies and a photographic presentation of medicinal plants; and includes a library, ethno-botanical and medicinal demonstration gardens and nurseries.

Medicinal Plants of the Local Area

***Rauwolfia Tetraphylla*:** Snake and scorpion bites are common in this area and are often highly poisonous. All parts of this plant are considered to be useful in curing snake (most) and scorpion bites. The leaves in particular are useful, being applied in dosages of about 10 grams after being ground. The fruits are edible and are considered to act as a preventive medicine for snake bites.

***Achyranthus Aspera*:** This plant is considered to be an important auspicious medicinal plant. All parts of the plant are medicinally useful. The roots can be used as a tooth brush to strengthen the teeth, cure toothache, to reduce swelling in the chin and to give freshness to the face. A preparation of the leaves (pasted with sesame oil) is believed to cure dog bites. Eating the roots (pasted with water) reduces the effects of scorpion and snake bites.

Leucas Aspera: The whole plant is considered to be medicinally useful and sacred. The flowers can be collected in the mornings (traditionally it is believed that plants have greater powers depending on when they are collected), boiled with water and then given as cough medicine for infants. The leaf juice is considered to be useful for many different types of skin troubles.

Tinospora Cordifolia: In Sanskrit, this plant is called Amruth. The word Amruth literally means immortal. It is a "wonder" plant known traditionally for its rejuvenating properties, particularly the leaves and stems. The liquid obtained from boiling the leaf can be used as a medicine for diabetes.

Continuing Work

1. Establishment of a community register that documents the rich traditional knowledge of a given village's use of bio-resources, their availability and conservation practices related to these resources. The objectives are as follows: to collect information with community involvement of school children, youth clubs and women's groups to share the local knowledge of bio-resources for mutual benefit with other communities in India to revitalise local health traditions and technologies and to nurture respect for local plant doctors who are being devalued in the face of powerful advertising of western medicines.
2. An essential aspect of the overall work is the creation and maintenance of havens in which the indigenous forest may develop naturally and serve as a gene bank. The strategies are as follows: to reintroduce individual species into their former habitats, to regenerate whole communities and to conserve rare and endangered species to protect 2,500 acres of remnant vegetation using local community support in collaboration with the Forestry Department.
3. Spreading the message to local people about the importance of the indigenous forest. For example, to encourage the understanding that all remaining areas of the TDEF be left undisturbed and to coordinate efforts

by communities that are aimed at the preservation and restoration of remnant forest.

4. Environmental education programmes are being developed, for example: developing curriculum for village schools and for adult education programmes using information about local bio-resources and traditional skills and technologies to empower and support local communities in running training programmes with the practitioners of local folk medicine.
5. To continue establishment of individual village gardens containing food and medicinal plants.
6. To establish Herbal Dispensaries in the villages which serve the local communities.

Through our outreach work with revitalising local medical knowledge and skills, together with protecting and re-establishing the bio-resources (in particular medicinal plants), we are endeavouring to establish a sub-culture that can survive within the wider reality of modern India. A sub-culture that can continue to seek and find its inspiration from the diversity of the biosphere and that has a firm enough base to sustain itself, that interfaces with the world as a whole and relates equally well to the traditional villagers who seek to maintain their lifestyles within the changed landscape of industrialised India.

The Seed Savers' Network

Access to good quality planting material is a top priority for the villagers, women's groups and agricultural teachers and students whom we recently trained in community seed banking in Tonga and the Solomon Islands. The seed production segments of our training were greeted with great enthusiasm as the concept is a relatively novel one to peoples who garden mostly vegetatively reproduced plants. Our Australian-based Seed Savers' Trust facilitates the collection, multiplication and public access of a wide range of local seeds and other planting material for food and agriculture. This, in our view, is the essential first step towards household food security and regional economic self reliance. It also aids the conservation of agricultural bio-diversity. Our educational organisation, The Seed Savers' Trust and its network of 4000 farmers and gardeners, has been actively conserving useful plants in Australia for the last ten years. In the Pacific we are currently setting up:

- The collection and dissemination of local vegetable varieties suitable to sustainable farming systems;
- Databases on the availability and location of important fruit and nut trees and other plants of economic importance such as for fibre, perfume, cosmetics, soil regeneration, fodder and forestry;
- Living collections of important food trees in schools, church grounds, markets, and other public spaces;
- Training in home garden design, sustainable farming techniques and the promotion of under-utilised plants, including permaculture practices; and
- A Pacific exchange network for seed banks and field collection.

Over the last year we have initiated seed and planting material networks in the Solomon Islands in collaboration with Appropriate Technology, Community and Environment (APACE); in Cuba, with the permaculture-oriented Australian Green Team and in Tonga with The United Kingdom Foundation for the Peoples of the South Pacific. In each case a team centred very much around indigenous NGOs was formed to administer the network with initially modest newsletters including lists of varieties offered. During recent Community Seed Bank training in Tonga and the Solomon Islands, we found that many plants significant to their cultures are endangered, such as:

- Local varieties of fruit and nut trees that were once widely grown have been the victims of cyclones, disease or development, and are not replaced, for example the much-loved Tongan native nut, the ai (*Canarium* spp.), or the national emblem, the fragrant Heilala tree (*Garcinia sessilis*), are rarely grown nowadays in backyards or plantations;
- In the Solomon Islands, small populations of vegetable varieties of local cucumbers, watermelons, Chinese

cabbage and open pollinated corn are disappearing through cross pollination with imported hybrids; and

- Many medicinal, fibre, and dye trees and shrubs are increasingly difficult to locate because of land clearing for agricultural exports.

The promotion of a greater range of food tree and lesser known tree crops grown for posterity in public places, will not only help redress the poor diet of urban children and adults but will also link conservation and utilisation of plant genetic resources.

At IPC6 I am hoping to make contact with others who are working in this field with a view to coordinating a global network of seed saving groups.

6

Maintaining Seed Viability and Vigour

At times, it is necessary to store seed for a period of several months to a year or more. Unfortunately, exposing seed to temperatures of 90°F or higher, or to relative humidities of 60% or greater, can lead to a rapid decline in seed germinability and vigour. As a rule of thumb, stored seed best maintains viability and vigour when the combination of air temperature in degrees Fahrenheit and percent relative humidity of the air is 100 or less. Thus, a seed storage relative humidity of 30% or less at a temperature of 70°F or lower would maintain viability and vigour in the seed of most plant types.

Conversely, as is often the case where seed is stored, air temperatures during part of the year may be 80° to 90°F or higher, and the relative humidity of the air may be 70% or greater. Such conditions, often found in the southern states, can lead to lower germinability and vigour in only a few months.

Seeds exposed to air gain or lose water according to the relative humidity of the surrounding air. At 50% atmospheric relative humidity, the equilibrium moisture content of wheat and rye seeds is about 12%; of barley, about 11%; and of oats, about 10.5%. Small-grain seeds and, in fact, seeds of most plant kinds store reasonably well if the moisture content in the seed is 10 to 12%. Seeds with a high fat or oil content, such as soybeans, store better at 9 to 10% moisture or lower. Oat seeds should be dried to 10% or less moisture content.

At 60% or greater atmospheric relative humidity, seeds of many plant kinds rapidly gain water. Thus, the equilibrium At 90% atmospheric relative humidity, the seed moisture content of the several smallgrain crops swells to 20 to 23%. Viability and vigour are lost rapidly under these conditions.

Effects of Moisture Content in Seeds

Seed moisture	*Condition*
4-8%	*Little or no insect activity (too dry for most insects).*
10-12%	*Satisfactory to store most seeds in open storage and in cloth bags or moisture-resistant containers.*
14-16%	*Molds (fungi) may grow on and in seeds in open storage and on seeds in cloth bags or sealed containers. Harmful to seeds of many plant kinds.*
18-20%	*Seed may heat because of seed respiration and microbial activity. Seed declines rapidly in viability and vigour.*
24-60%	*Seeds may rot.*
40-60%	*Germination begins.*

Although it usually is not practical to control the temperature and relative humidity of the space where carryover seed is stored, you should pick a place where the temperatures are below 86°F most of the time and where the seed is not exposed to humid conditions.

Be sure the seed is dry when stored. Seed of oats, wheat, barley, rye, soybeans, red clover, and timothy often is farm grown. Such seed may be harvested at moisture contents of 16 to 18% or higher. This seed should be dried carefully to 10 to 12% seed moisture content soon after harvesting. Do not dry such seed at temperatures over 90° to 95°F, and do not maintain elevated temperatures any longer than is required to dry the seed.

Seed of corn, sorghum, alfalfa, and most grasses usually is not farm grown under Pennsylvania conditions. To maintain viability and vigour in purchased carryover seed of such plant kinds, reseal the containers to reduce air interchange, and store where temperatures are moderate and atmospheric relative humidity is as low as possible.

At times, farm-grown red clover or timothy seed is bagged and laid in a bin on top of freshly harvested oats. Moisture migrates upward as the oats dry and may condense on the surface of or in the bags of timothy or red clover. This can lead to a rapid decline in germination and vigour. Remember that heat and humidity are enemies of seed viability and vigour. If insects are a problem in stored seed, see the following.

Controlling Pests in Stored Grains

Managing grain in storage to prevent insect damage is just as important, or more so, than managing the crop while it is growing in the field. When insect infestations develop, they can reduce grain weight by up to 10%. A farmer with a grain bin containing 30,000 bushels of corn valued at $3.00 per bushel would lose $9,000 if an insect infestation developed and consumed 10% of the grain mass. This loss does not include dockage or the cost of eliminating the insects from the grain.

To prevent losses from insect feeding, growers should implement a sound integrated pest management (IPM) programme in their grain bins. An integrated pest management programme involves: (1) grain bin and harvest equipment sanitation, (2) proper harvest equipment adjustment, (3) grain cleaning, (4) uniform distribution of the grain during uploading into the bin, (5) moisture management, (6) temperature management, (7) periodic grain monitoring, (8) residual insecticide treatment for long-term storage, and (9) fumigation as a last resort. If the first eight tactics are used effectively, fumigation should be needed only rarely.

Preparing the grain bin to receive grain is one of the most important steps in an IPM programme. Before placing new grain in a bin, remove all old grain, which can act as a potential source of insect populations that will infest the new grain rapidly. Once you have removed the old grain, clean the bin. Sweep or wash off the walls and floor, and be sure to remove grain from any ledges above the door or between the ceiling and walls. Use a shop vacuum to remove debris from under metal flashing at the floor/wall interface and other tight areas that cannot be reached with a broom or hose. It is a good idea to remove the false floor and grain auger once every five years or so and clean out any grain residue and dirt.

Remember that the insects that infest grain range from 1/16 to about ½ inch in length. Large numbers of these small insects can be harbored in very small amounts of debris. If half of an initial population of 1,000 insects is female, and if each female produces 200 offspring, the population can produce approximately 10 million insects in two generations. At a grain temperature of 90°F, many species produce one generation per month. If a grain handler detects live insects in your grain, you may be docked or sent home with the grain to eliminate the infestation. A clean grain bin is a good start toward preventing insect injury.

After cleaning the interior of the grain bin, remove old grain that has collected around the outside of the bin. To kill any insects that were in the grain, burn or bury it. Also remove or mow any weeds that have grown up around the outside of the bin. Because many grain-infesting insects are seed feeders, they can develop on the seed of these weeds and then move into the grain. Weedy areas also can harbour rodents that can enter grain storage structures, especially wooden ones. After cleaning, inspect all areas of the bin for leaks or gaps where insects could enter. Repair any cracks and continue to monitor the facility for damage.

Once you have prepared the grain bin, make sure that your harvest equipment is cleaned and properly adjusted. A corn picker or grain combine that was used last year and then parked in the equipment shed without cleaning can harbour insect infestations at the lower ends of augers or on screens inside the combine. Remove plant residue from all augers and other areas within the combine before using the equipment to harvest grain. It also is important that a combine be adjusted properly, to prevent shattering of the grain. Small pieces of grain make it easy for a number of grain-infesting insects to become established and develop; most insect species have a difficult time feeding on whole kernels of grain.

Grain harvested with a clean and properly adjusted combine also should be run through a grain cleaner to further remove fine materials on which insects can feed. Grain with many broken kernels also can be run through a cleaner, but this will lead to additional weight loss. Clean, whole grain helps in insect management, but also is important in proper aeration of the grain mass during drying and storage. If fine materials are not present, air drawn into the grain from outside the bin will move more uniformly through the mass. A grain distributor is extremely helpful in preventing aggregations of fine material by spreading it uniformly across the grain mass. When a distributor is not used, the fine material will concentrate in a column at the centre of the grain mass. Because this column is more dense, air will move around it through the grain mass and allow moisture to build up in the centre of the bin. Moisture buildup is the first step in grain quality deterioration.

Lowering the grain temperature and moisture levels will help inhibit fungi, bacteria, and insect feeding and reproduction. Thus, harvesting and storing grains during cool, dry periods will help prevent pest infestation. Reduce the grain moisture to 13% or lower while loading or after the grain has been loaded into the bin. Level the grain surface and remove any peaks. As with a fine material buildup, the peak in the grain does not dry properly because air moves through the grain mass below it. Once you have loaded the clean, dry grain into the bin and levelled off the top, the next step is to cool the grain after outside temperatures fall below 40°F in late fall. Lower the grain mass temperature by running the fan during periods when the temperature is below 40°F. If possible, do not run the fan if the outside air is moist; wait until the air is dry. The amount of time necessary to cool the entire grain mass varies depending on the fan capacity, the initial bin temperature, and the height of the bin. Do not cool the grain below 32°F. An airflow rate of 1/10 cubic foot of air per minute per bushel usually is ample to cool dry grain.

After the grain has been prepared properly for storage, its quality still can decline if proper monitoring does not occur. During the winter, check grain for moisture and temperature buildup once every two months. During the summer, check grain once a month. Changes in grain moisture content and temperature are the first signs that grain quality is declining. Moisture migration can occur in a bin when outside temperatures are greater or lower than the grain mass temperature. When the outside temperature is cooler than the grain temperature, air in the grain mass drops along the bin walls and then rises in the centre of the warmer grain mass. As the air rises, it picks up moisture and drops it in the centre at the top of the grain. When the outside temperature is warmer than the grain mass, the air circulates in the opposite direction and deposits moisture along the bin wall at the top of the grain mass.

As the moisture level increases, fungi and bacteria begin to grow as they feed on the grain. Through respiration, these organisms give off heat. This cycle then continues to drive up the grain temperature, which in turn increases the moisture levels. Once the temperature rises above 40° to 50°F, insects can begin to develop and reproduce. The higher the temperature, the faster the insects complete each generation and increase in numbers. This cycle will continue until the grain is dried and the temperature is brought back down using the fan.

Once grain crusting occurs, drying efficiency is reduced because the air will not move through this dense area in the grain mass. It may be necessary to break up the crusted area. Prevent this cycle from getting out of control by monitoring the grain periodically and using the fan as your primary management tool.

If you will be storing the grain for only 6 to 9 months, these measures should be sufficient to prevent insect infestations. If you will be storing the grain for more than 9 months, however, you may need a residual insecticide treatment. Cooled grain will maintain temperatures below 60°F well into the summer, but by late summer temperatures may rise to levels at which active insect populations will colonize the grain and build up population levels. The longer grain is stored, the greater the chances are that insect infestations will develop. When possible, do not store grain for more than one year. The best residual insecticide treatments provide only 18 months of protection.

Preventing Infestations

Good sanitation and local environment maintenance are generally sufficient to prevent major pest outbreaks. However, if pests have previously been a problem or grain will be stored for an extended period, other options are available to avoid future problems. There are four main modes of pesticide application for stored grain: empty bin sprays, grain protectants, top dressing/pest strips, and fumigation.

After bins have been cleaned, an empty-bin treatment may be applied. Pay special attention to cracks, crevices, and "hidden" or out-of-the-way spaces. Bins should be sprayed at least 2-3 weeks before filling with grain; allow 24 hours for sprays to dry.

Grain protectants are insecticides added to grain as the bin is being filled. These products kill insect pests as they crawl or feed on treated grain. Grain protectant treatments are recommended for grain in extended storage, storage in flat structures, or under conditions favourable to pest development. Protectants may take the form of liquids or dusts; they are generally applied during auguring or loading. Chemical formulations should not be applied before high-temperature drying. If using these products, keep bin temperature and humidity levels low. Heat and moisture shorten the residual life and efficacy of most chemical grain protectants.

Moth pests often enter storage areas from the top surface, so some grain protectants may be applied as a surface treatment. Bacillus thuringiensis (Bt) is a highly selective biological insecticide that provides protection against moth larvae. This may be applied to the surface, or another treatment (Actellic 5E, pyrethrins + piperonyl butoxide, or diatomaceous earth) may be used. Top dressing treatments should be mixed into the top 4 inches of the grain.

If the surface is already infested with moth larvae (webbing is present), remove the webbing and rake the surface treatment to a depth of 5 inches. Resin strips (dichlorvos or DDVP) can also be hung in the top air space; these strips help control adult moths. Use one strip per 1000 cubic feet of air space, and replace it after 3 months. When using strips, the bin top and roof vent must be temporarily sealed (aeration will make this treatment ineffective). Before re-aerating, remember to open the roof vent.

Using Fumigants

The cheapest and safest method of controlling grain pests is prevention. Consider fumigation only as an emergency treatment. Once grain has become infested with insects, the quickest and most effective way to remedy the problem usually is with a fumigant. All fumigants, however, are hazardous and must be handled with extreme caution. Fumigants are considered restricted-use pesticides, and their container labels all carry the signal word "Danger" and a skull-and-crossbones symbol. Only methyl bromide, chloropicrin, aluminum phosphide, and magnesium phosphide are available for use as grain fumigants. As of 1990, fumigation of grain storage facilities requires special certification. Contact your local Penn State Cooperative Extension office or the Pennsylvania Department of Agriculture for details. The use of methyl bromide has also been gradually phased out, and its registration will end in 2005.

Although it is legal to acquire private certification to apply fumigants on the farm, a farmer should consider seriously the hazards associated with fumigation products before attempting to fumigate grain. The most commonly used grain fumigate is phosphine gas produced by aluminum phosphide or magnesium phosphide coming in contact with water. These materials are sold as pellets or tablets that are placed in the grain mass. When they make contact with normal air moisture levels, they slowly begin to liberate the phosphine gas. The amount of time that an applicator has to place these tablets in the grain mass and get out before phosphine levels become toxic depends on moisture and air temperature levels. If a tablet or pellet contacts too much moisture, the reaction to liberate the gas is rapid, resulting in an explosion. This explosion can harm the applicator and set the grain mass on fire.

In addition to this significant hazard, the fumigant is a very small molecule that is only slightly heavier than air. Because of these properties, the gas is hard to contain unless the grain bin is tightly sealed. Phosphine can move through plastic, concrete, wood, and any crack or seam in a metal bin. Because it is heavier than air, it will slowly displace air in low areas and become concentrated if it escapes from the bin. Therefore, any work areas, nearby homes, or livestock shelters near a bin under fumigation can build up concentrations of the gas high enough to kill livestock or humans. Properly handling and using fumigants requires considerable experience. Novices can easily harm themselves, employees, family members, and livestock. For this reason, fumigation is best left in the hands of experienced professionals.

For those who choose to use fumigants, the methods of application, dosages, temperature, moisture, exposure periods, and all precautions are listed on the label. Be certain to read and understand all directions before using any fumigant. In addition, consider the following recommendations:

- Fumigants generally are not fully effective at temperatures below 60°F.
- Aluminum phosphide is less likely to impair seed viability than are other fumigants.
- Aluminum phosphide and magnesium phosphide are extremely corrosive to gold and copper. As a result, electrical and computer circuitry will be damaged if it contacts the gas.
- Never use methyl bromide on any grain intended for seed use.

Safer Options

Producer and consumer safety are important considerations when making decisions about insect management. Choosing low-toxic control options will help keep you safe and increase consumer confidence in your product. Diatomaceous earth (Insecto, Protect-It, etc.) is one such product that can be used as a top-dressing. This substance is milled from the silica-rich cell walls of fossilized sea algae. As insects crawl through the grain, their hard cuticle coverings are abraded by the sharp edges of the diatomaceous earth. The insects then desiccate and die. Diatomaceous earth should be used to treat the first loads in the bottom of the bin and the last loads in the top. As with any treatment, use proper protection and caution; the product can be irritating to workers and it can abrade equipment.

Bacillus thuringiensis (Bt) provides highly selective control for moth larvae. Bt will not control weevils or other beetle pests. However, if the Bt gene was genetically engineered into the grain crop to be stored, the grain may already be well protected against larval moth damage. Many newer crops express Bt in the grain itself, as well as in other parts of the plant. No conclusive research has been done on the effectiveness of Bt grain in stored conditions, however, so continue to sample grain for pest presence.

Carbon dioxide can also be used as a safer fumigation alternative, but the bin must be completely sealed. While the bin is sealed and fumigated over several days, the carbon dioxide keeps the insects' respiratory openings open; the pests then dehydrate. Heating or cooling the bin to extreme temperatures may also kill some species. Biological control using beneficial natural enemies is another option; some companies sell species that attack maize weevil, granary weevil, confused flour beetle, lesser grain borer, sawtooth grain beetle, and Indian meal moth.

Eliminating unnecessary chemical applications improves food quality and saves money. Be sure to properly identify insect pests; many are very small and have only subtle characteristics. Some fungus-eating beetles can appear to be pests of the grain itself. However, their presence indicates a different problem: high moisture levels. To be certain of the identity of a stored-grain pest, consult a professional opinion.

Some insects are becoming resistant to various chemicals, most notably malathion. A few populations of insects (Indian meal moths, red flour beetles, lesser grain borers) are resistant to malathion control, so treatment is ineffective. Although they are labelled as grain protectants, malathion dusts are not recommended. Insect resistance is widespread and the public is concerned about malathion residues in food, so its use is discouraged.

Fungus Beetle Infestations in Stored Shelled Corn

Several species of fungus beetles, including the foreign grain beetle, have increased in abundance in the past few years. These insects do little damage to the grain. They simply use the grain as a home, becoming most obvious in late summer and fall. Grain contamination from these insects can result in dockage, however, when the grain is sold.

The presence of fungus beetles in grain indicates a moisture and heating problem. Under these conditions, fungi grow and the beetles feed on them. Two things can be done to keep these beetles out of storages: eliminate fines, debris, screenings, etc., from the storage during filling, and keep the storage amply ventilated.

Indian Meal Moth Infestations

Over the past several years, an increase of Indian meal moth damage has been noted in farm storages where grain has been held over winter. Indian meal moth is not a common pest of stored grain in the state, but it is a serious pest in those relatively few storages where infestations develop. At least two ways of attacking a stored grain Indian meal moth infestation exist: fumigate as previously discussed, and use a grain surface treatment of Dipel. Dipel 2X is a formulation of bacteria that is a pathogen to the larvae when eaten. Mix ½ pound of Dipel 2X in 10 gallons of water for each 500 square feet of grain surface; sprinkle over grain surface and incorporate into the top 4 inches of grain.

Corn

Corn's high carbohydrate content makes it a high-energy feed. An average sample of corn grain contains 70.5 percent carbohydrate (starch), 9.6 percent protein, 0.33 percent P, 0.41 percent K and 4.4 percent oil. Approximately 56 percent of the corn grown in the United States is fed to livestock, 18 percent is exported, and 13 percent is made into ethanol. Most of the exported corn is used for livestock feed. The remaining 13 percent of the crop is used for food, alcohol, or industrial purposes.

Types: Yellow and white dent, waxy, high-lysine, high-oil, sweet, and pop are examples of contemporary corn types. Each has unique characteristics that affect its use and, in some cases, how it is grown.

Yellow Dent Corn

Most of the corn crop in the United States is yellow dent. Its vitamin A content, high feed value, and availability of adapted superior hybrids account for its extensive use. Of the cereal grains, it contains the highest content of carotene (Vitamin A). Dent corn originated from crosses of flint and floury corns. Dent hybrids vary in the proportion of hard and soft endosperm.

White Dent Corn

White corn types are equal to yellow types in carbohydrate content but are deficient in vitamin A. White types are used primarily for human foods, breakfast cereals, and grits. Almost all white types are grown under contract; therefore, it is advisable to have a market before growing. Contracts for growing white dent usually offer a premium over the yellow dent price. White hybrids in maturities adapted to Pennsylvania are scarce, however, and their yields may not be competitive with yellow dent yields. A standard yellow dent check hybrid produced 20percent more grain yield than the best white hybrids in a Penn State yield trial. Yield data should be a prime concern of the producer before growing white corn.

White corns must be grown in isolation from yellow dents to prevent cross-pollination. A mixture of white and yellow downgrades either type at the market.

Waxy Corn

The carbohydrate or starch granules of regular dent corn consist of approximately 75 percent amylopectin and 25 percent amylose. Waxy corn has nearly 100 percent amylopectin.

Waxy corn initially was recognized as a valuable source of industrial starch. Results of the numerous feeding trials and evaluations of waxy corn that have been conducted over the last 20 years occasionally have shown a benefit in feeding waxy corn, but they have not been consistent. In general, changes in feed efficiency and production have been small and not significant, but usually in favour of the waxy corn. In one summary of nine beef feeding trials, the effect of waxy corn on animal gains ranged from a decrease of 3.3 percent to an increase of 10 percent. Averaged over all nine, waxy corn had a 2.2 percent advantage.

The Miner Institute in Chazy, New York, recently conducted a feeding trial to compare waxy and nonwaxy corn for dairy cows. It found no significant differences between the waxy and normal corn in milk production, butterfat, and dry matter intake. Similar results have been documented for the feeding value of waxy versus normal corn silage.

Yields of some waxy corns have been comparable to those of normal dent corn. Growers should have some estimate of the relative yield performance of specific waxy hybrids before growing any substantial acreage. Waxy corns must be grown in isolation from other corn types in order to maintain purity standards for industrial use. Keeping the harvested grain separated by 12 to 16 border rows from the rest of the field is considered adequate isolation; 5 percent contamination with regular corn pollen generally is considered the upper limit.

High-lysine Corn

All protein is a combination of building blocks called amino acids. Although normal dent corn contains about 9.4 percent protein, the quantity of two essential amino acids, lysine and tryptophane, is below nutritional requirements for nonruminant animals and humans. High-lysine (opaque-2) corn corrects this deficiency and may be advantageous in swine feeding rations. High-lysine corn also has virtually no hard or vitreous starch in the kernel. Adapted varieties of high-lysine corn are limited in number and generally are lower in grain yield than normal dent corn varieties. The softer kernels of lysine corn break and crack more often at harvest, which can lead to higher incidences of kernel or ear rot. Because cross-pollination results in normal dent corn, high-lysine corn production fields must be isolated from other corn fields. A separation distance of 100 yards is suggested.

High-oil Corn

While normal dent corn contains 3.5 to 4.5 percent oil, commercial high-oil types contain as much as 1.5 times this amount. Commercial high-oil corn hybrids have not been widely used because their yield potential is lower than that of normal hybrids. Recently, however, an alternative system for producing high-oil corn has been developed. It involves planting a blend consisting of about 92 percent of an adapted but male-sterile hybrid and about 8 percent of a special high-oil pollinator line. The pollen shed by these pollinator plants carries a special gene that causes a kernel to produce a much larger germ or embryo. Pollen from these plants pollinates the entire field, resulting in higher oil and protein in the grain on adapted hybrid plants.

This system of producing high-oil corn was developed by DuPont and is known as the Topcross system. Because the yield of the pollinator plants is low and pollination is dependent on 8 percent of the plants, some concern exists about the yield potential of these blends compared to normal fertile hybrids. Initial testing has shown that yields of the high-oil blends are 90 to 100 percent of those of their normal counterparts. Testing these blends is difficult, since they have to be isolated from normal corn and thus cannot be tested in conventional strip tests or small plot evaluations. Grain from these blends contains about 7 to 7.5 percent oil compared to 3.5 to 4 percent for normal corn; slightly more protein, lysine, and metabolizable energy than normal corn; and less starch than normal corn.

Brown Midrib Corn

Brown midrib plants are characterized by a brown pigmentation in the leaf midrib at the V4 to V6 stage and a light brown coloration of the pith after tasselling. Brown midrib hybrid contains a gene (bm3) that causes a lower lignin content in corn plant tissue. Lignin is a relatively indigestible compound that limits the digestibility of the fiber in the corn plant. The reduced lignin in brown midrib corn results in silage with fiber that is more digestible than normal corn.

Animal feeding trials have shown about 10 percent greater intake and increased mild production with brown midrib silage, as compared to normal silage. Brown midrib lines frequently have been low yielding, however, and have had poor standability if left for grain production. Although recently released brown midrib hybrids have improved these features, they still are lower than those of comparable normal hybrids. In our trials, brown midrib lines have yielded about 10 to 15 percent lower than conventional hybrids in fields where they are best adapted, but have yielded 15 to 25 percent less than conventional hybrids where European corn borers, gray leaf spot and other pest problems are present. When considering a brown midrib corn planting, growers should weigh the potential for improved milk or animal gains, which can be significant, against the higher production costs and lower yields per acre.

Popcorn: Popcorns are essentially small-kerneled flint corns and are among the most primitive of the surviving races of corn. Kernels contain a very hard endosperm with only a small portion of soft starch. Popcorn hybrids usually yield less than half of normal dent hybrids. Maximum quality is achieved by minimizing mechanical damage and drying with low heat to a moisture of 13.5 percent. Overdrying and kernel damage result in reduced popping volume. Most popcorn is produced under contract.

Open Pollinated: The availability and performance of superior hybrids essentially have made the practical use of open-pollinated corns obsolete. Open pollinated corns often suffer from low yields and poor standability. Four open-pollinated varieties-Early Butter, Reids Yellow Dent, Golden Surprise, and Lancaster Surecrop-were evaluated in Pennsylvania Hybrid Corn Tests. Of 59 county tests, the average yield for the open-pollinated varieties was 82.9 bushels per acre with 29.4 percent down stalks. The average yield of the hybrids was 132.9 bushels per acre with 7.8 percent down stalks-a 50-bushel or 60 percent yield advantage.

Hybrids: Hybrids are available for all types of corn, and include double crosses, three-way crosses, single crosses, and modified crosses (or sister line crosses). These terms simply describe the number of inbred parent lines used in producing a hybrid and do not guarantee that one is superior to another. The merits of a hybrid rest with its genetic potential to produce grain and/or silage.

The performance of a particular hybrid may vary from location to location and from year to year. Silking dates for the hybrids being grown on a particular farm will vary, and hybrid maturity differences will increase on farms that frequently encounter drought stress. This variability has two implications for corn growers. First, you should use more than one hybrid each year. Reserve part of your acreage for evaluating new hybrids, but grow the bulk of your acreage to hybrids with which you have experience and that have a known performance record on your farm.

The second implication of the variability of hybrid performance is that hybrids should be selected on the basis of performance in a number of environments. Average performance over a wide range of environments is usually the best predictor of hybrid performance on your farm next year. As a result, two-year performance averages or average performance results from several test locations within a year increase the reliability of data on hybrid performance.

Hybrids with wide ranges of adaptability have less variation and consistently rank as superior. Although specific hybrids may adapt particularly well to some environments, these environments often are difficult to identify and usually are found after growers have had several years' experience growing a specific hybrid.

Male-sterile corn, grown for high stalk-sugar content, produces no grain and usually yields less dry matter than adapted grain-producing corn hybrids. This limits flexibility in terms of harvesting the crop for grain if needed.

To ensure a good, uniform stand, plant high-quality seed. Certified seed or seed bearing the brand tags of a reputable dealer is the best assurance of quality. Most of the seed corn is tested for germination and vigour and is of high quality.

Varieties

Hybrid Maturity: Along with yield potential, maturity is of primary importance in selecting a hybrid that will produce mature, high test weight grain during a normal growing season for a particular area. Relative maturity among hybrids in performance reports is best shown by grain moisture content. Hybrids are ranked from earliest to latest in each of the test reports. Days to maturity and growing degree days (GDD) are two methods of expressing the energy required for corn to mature.

The relative maturity rating of hybrids is just that, a relative value; an 85-day hybrid is not ready to harvest in 85 days. One can be confident, however, that an 85-day hybrid will reach a particular moisture content earlier than a 95-day hybrid. Once a grower establishes the rating for his or her farm, most hybrids of similar relative maturity should mature. Maturity ratings may vary slightly among seed corn companies. The best way to compare the maturity of hybrids from different companies is to compare grain moisture contents of the two hybrids in the same trial.

Growing degree days are a daily accumulation of heat for crop growth. Corn does not grow when temperatures are below 50°F, and temperatures above 86°F do not increase plant growth rate. Most seed-corn suppliers include GDD ratings on their seed tags or in their hybrid descriptions. The hybrid GDD ratings are determined from planting to black layer formation. The formula for calculating growing degree days is to add the daily high temperature (86°F maximum) and daily low temperature (50°F minimum); divide this figure by 2 and subtract 50. The remainder represents the heat units for one day:

GDD = (daily high + daily low) ÷ 2 - 50

Example:

84 (daily high) + 62 (daily low) = 146

146 ÷ 2 = 73

73 - 50 = 23 GDD

A minimum-maximum thermometer enables producers to determine and record growing degree days for their farms. We have found that GDD ratings vary somewhat for hybrids depending on where the rating was done. Hybrids typically use about 200 fewer GDDs to mature in Pennsylvania compared to some mid-western locations, so companies that use GDD ratings based on midwest data may overestimate the GDD requirements of their hybrids in the east. Our data indicate that a 105-day hybrid matures in about 2,350 GDDs when grown in Pennsylvania, despite the fact that some 105-day hybrids are rated at 2,600 GDDs. Hybrids with similar GDD ratings may vary in moisture content at harvest because of different drydown rates following physiological maturity.

The Pennsylvania corn maturity zones are based on longterm weather data, including frost-free growing days and growing degree days.

Bulbs

Bulbs which grow and flower well in the hills include Agapanthus, Anemone, Cylamen, Eurycles, Fritillaria, Iris, Ixia, Hyacinth, Moraea, Montbretia, Nerine, Paeonia, Kniphofia, Ornithogalum, Ranunculus, Tulip, Watsonia, Sparaxis, Zantedeschia Tuberous Begonia, etc. Although some hardy types of Freesia, Daffodil, Narcissus, Gloxinia, and Lilium may flower in plains but they do not show very attractive display of colour and often fail to flower in the second year.

Planting

Bulbous plants are grown in beds, shrubbery, herbaceous border, and grassy land and in pots and bowl. Bulbs, corms, rhizome or tuber are usually planted when they have shown signs of sprouting after dormancy. In Canna, however, the rhizomes are taken out of bed in growing condition and replanted after a week or so. The planting materials are normally placed deep in the soil and the soil around it is gently pressed. The depth of planting varies with the type of plants and size of bulbs. For most of the bulbous plants grown in tropical garden, the planting depth is between 3-10cm. Bulbs are used for vegetative propagation. Various methods and structures used for propagation are as follows:

1. Division of clump-Rhizomatous plants like Canna, Alpinia can be divided into small clumps.
2. Offsets produced laterally are separated for multiplication.
3. Cormlets or bulblets produce new plants.
4. Scales of Lilium are separated and used for propagation.

5. Bulbil arising from the axils of leaves produce new plants.
6. Pieces of bulb disc with leaves or rhizome also develop new plants.

Soil

Soil should be sandy loam or loamy sand in texture for better production of bulbs. It should be moderately fertile with good water holding capacity and with pH ranging between 7.0 - 8.0. For pot culture the ideal pot mixture should contain 2 parts sand: 2 parts leaf mould: 2 parts well rotten farmyard manure: 1 part soil and ½ part charcoal. Bulbs prefer loam or sandy loam soil. In stiff clay, rooting is delayed and too much moisture often causes rotting. If the soil is not perfectly well-drained, the bulbs may be planted on a bed of sand. Watering is not required after planting as the fleshy underground stems contain sufficient food materials to develop initial growth of root and shoot. Before root formation, watering proves injurious to the bulbs and helps in rotting.

Climate

In general, cool weather, high atmospheric humidity and moist soil conditions are congenial for the manufacturing of food, better production of high quality flowers and bulbs. According to climatic requirement of these bulbs, they are grouped into two categories i.e. Warm climate-Gladious, Narcissus, Daffodils, Freesia, etc. Modified environmental conditions are to be provided during winter and summer seasons for the production of high quality blooms. Most of these plants prefer sunny situation, while Eucharis, Zephyranthes thrive better in semishade; few types, e.g. Pancratium, Haemanthus, Zephyranthes grow and flower well in sun and semishade. Bulbs, corms and tubers should not come in direct contact with fresh organic manure, rhizomes are not usually affected.

Manuring

Well rotten farmyard manure @5kg/sq m should be incorporated at the time of soil preparation. The application of 20g each of P_2O_5 and K_2O/sq m has also been found beneficial at the time of planting bulbs. Nitrogen @40g/sq m into two-three splits should be applied.

Intercultural Operations

Weeding and Hoeing: Weeds should not be allowed to grow and should be removed as soon as they appear otherwise they compete for nutrients and water with the main crop. Hoeing helps in aeration of soil and better development of bulbs. In early stages of growth frequent weeding and hoeing are required.

Irrigation

Liberal application of water at all stages of development is highly beneficial. Generally 7-10 days interval is appropriate in winter whereas in summer it should be about 5-7 days.

Staking

Bulbous plants like Gladiolus, Dahlia, Lilies, etc. produce flowers on long stem and thus need support. Otherwise spiks may bend down and break with strong winds. Bamboo stakes can be provided.

Harvesting of Flowers

The stage of harvesting of flowers depends upon the distance of travel. It varies with the kind of flower. For local market, gladiolus and tuberose should be cut when lower 2-3 florets have opened. Whereas for distant market, flowers are cut when basal florets show colour. Narcissus and Daffodil are cut at goose neck stage. Before harvesting the flowers, plants should be irrigated.

Lifting of Bulbs and their Storage

Tender bulbs like Gladiolus, Dahlia, Narcissus, Daffodils, etc. are to be dug from the soil 10-12 weeks after flowering has been finished. Before lifting the bulbs, water is withheld. After digging, bulbs are dried in shade for few days. Then these are treated with 0.2% Bavistin solution for 30 minutes and thereafter stored. Hardy bulbs are separated after 2-3 years and are again planted in the planting season.

Time of planting depends on the season of flowering, environmental condition of the region and condition of the planting material. In the plains, Gladiolus is planted from

September to November, whereas in the hills, the planting season is normally from April-June. Amaryllis, Hippeastrum, Harmanthus are planted in January-February to obtain flowers in March-April. Caladium and Canna are put in the ground in May.

Lifting and Storage

After the above ground portion dries out the bulbs are lifted out carefully. The root and shoot are cut and the bulbs are cleaned before storage. They should be stored in dark, dry and airy place until the time of next planting. Storage condition is very important for obtaining healthy bulbs. If the humidity is very high, fungus develops on the bulbs and causes rotting. In a very dry atmosphere the bulbs shrivel and may lose viability. It is necessary to lift all types of bulbous plants every year. Amaryllis, tuberose, Caladium, Pancratium, Zephyranthes may remain in the ground for three years before they are lifted. But water should be withheld when the bulbs are lying in dormant condition in ground or pot. Gladiolus, on the other hand, has to be lifted out every year after drying of leaves.

Roots—Tubers

The tropical origin root crops are cultivated in different climatic zones of the sub-tropics because of the fact that it needs shorter growth periods like potato and sweet potato and capable for tuberization in longer day lengths and lower temperatures. Daylength is also important but it varies between 6 to 12 hours. The short day conditions are highly favourable for economic yield of vegetables and tuber crops. The different varieties are suitable for different day length and temperature. However, adaptability in different micro-climatological regions needs suitable varieties for different environmental conditions and capacity for growing in different ecosystems. Thus physiological characters play an important role in adaptation. The day length plays an important role in tuberization of root crops. To secure good yield for Potato a day length of 12-13 hours is needed and night temperature 20°C increase tuberization. With short day length and temperatures sweet potato can be cultivated in sub-tropics round the year.

Soil water is necessary for germination and establishment of crops and this applies to root crops where vegetative propagation in used. The Yams also has a high correlation between rainfall, development of vine and final yield. The sweet potato needs 50-100 cm rainfall per year and yield is reduced even when drought occurs for a month or so but water logging is also not desirable. Tapioca however withstands low rainfall of 30-50 cm. Potato needs 60-75cm but it must be evenly distributed every week during growth and development period for better yield. It has been shown by research workers that the photosynthesis was increased with increase of CO_2 concentration and photosynthesis was closely related to the potassium content of the leaves and maximum photosynthesis occurs when 4% or more potassium are available. Similarly with high leaf area photosynthesis is more.

The yield of root crops grown during summer season in sub-tropical condition is generally lower due to long days and high temperature and also the high night temperature and as a result there is slow tuber initiation and growth while high humidity increases pests and diseases. But during winter all the conditions are favourable for higher yield. It is well known that nitrogen increases dry matter by increasing leaf area and thus there is a relationship between nitrogen content and leaf area. But in root crops high nitrogen increases top growth and leaf area, while net assimilation decreases. Root yield depends mostly on potassium. High potassium does not change leaf area index but increases net assimilation rate and thus high dry mater. In short, the favourable conditions for higher yield of tropical root crops in the sub-tropical conditions are favourable nutrient levels, light condition, photosynthetic activity, adaptability and soil conditions for endorsement of tuberous roots.

Beans

The beans belong to the family Leguminosae. They are considered as nutritious vegetables as they contain high amount of vegetable proteins. Besides carbohydrates and vitamins. The green tender pods are used as vegetable and also the dry seeds. The beans for drying are grown till the seeds are large enough and pod begins to dry. The pods are shelled and seeds are separated. The shelled dry bean has good market. The various beans cultivated in India particularly in home gardens and some commercially are as follows:

1. French beans (Phaseolus vulgaris)
2. Cow pea (Vigna sinensis)
3. Cluster bean (Cyamopsis tetragonoloba)
4. Broad bean (Vicia faba)
5. Lima bean (Phaseolus lunatus)
6. Soya bean (Glucine max)
7. Garden beans (Dolichos lab lab).

There are also many other beans, which are grown in south and northeastern India and in the hilly regions.

French Beans (Phaseolus vulgaris)

The French beans which are also called as snap bean has dwarf type and pole type. Generally it needs cool climate and therefore cultivated during Rabi season, cut as the bean is day-neutral plant. It can be cultivated in August-September and also from December-February in northern India. The pole type is a climbing plant and needs some support to climb, either bamboo or wooden sticks. The dwarf type is bushy and needs no support. The beans are planted in rows by hand or by a seed drill. In Rabi sowing if the temperature is low or cold germination may not be satisfactory. Bean should be harvested during tender stage or otherwise they develop strings. For dry beans it should be allowed to mature fully. The beans can be grown in a wide range of soil types but sandy loam is best.

Cow Pea (Vigna sinensis)

The cowpea is a warm season bean and grown during rainy season but the temperature should not be below 20°C. It is trailing plant and needs staking. The early varieties require 70 to 90 days to mature while the late varieties need 120 to 150 days. Well-drained loamy soil is best but can be grown in any type of soil. There are many cowpea varieties, some are bushy and dwarf and some creeping. The seeds also differ in colour, size, shape and pods. Many excellent improved high yielding varieties have been evolved by I.A.R.I. New Delhi.

Similarly, improved varieties have been evolved in Coimbatore, Punjab, Madhya Pradesh, Maharashtra etc. The early varieties can be harvested within 60 days while the late varieties in 90 to 100 days. To use as vegetable it should be harvested early otherwise soon they become leathery and stringy. Usually the pods are best for edible in tender stage. The dried seeds are used as pulse. The varieties, manure and fertilizer requirements, spacing etc. are mentioned in specific chapters.

Dolichos Beans

The Dolichos beans are grown throughout India. There are many local types particularly in eastern India and in south. It needs cool climate but can be grown in any kind of soils. The colour of flowers, size, shape and type of pods, size and shape differs from strain to strain. It is a good bean for kitchen garden. The tender pods are used as vegetable. Some plants are bushy and semi-erect while others are climbing. The flowers may be white, pink, purple and light violet and borne as axillary racemes. The pods may be flat or bulged with different shapes and colour. Some are improved by I.A.R.I. and at Coimbatore and Kalianpur. The seeds are sown in pits during June-July and climbing type needs overhead trellis or bower to trail. Tender pods are used as vegetable.

Cluster Bean (Cyamopsis tetragonoloba)

The cluster bean is mainly grown in the dry, warm, arid regions of India, which is known as Guar. The plant is hardy annual 1 to 3 meters in height. There are dwarf and tall types. The seeds are black, brown and white. The fleshy green tender pods are used as vegetables. The rough skinned hairy types are used as fodder for the cattle and also as green manuring crop. The crushed seeds are turned into flour known as guar gum, which is used in various industries like paper, textile and food and also it, has export market. In south and eastern India, it can be grown round the year but in arid zones of Rajasthan, Gujarat and Punjab it is grown commercially as it can thrive in hot winds, warm temperature and drought conditions. It can be grown in alkaline sandy loam soil. There are many good high yielding varieties and strains evolved by IARI and other Research Institutes.

Lima Beans (Phaseolus lunatus)

Lima beans are not commonly grown in India but it has importance in foreign countries. The beans are flat seeded and large and mainly it has two types, Pole type and Bush type. Lima needs mild cool climate and prefer heavy clay soils. The plants are deep rooted and the climbing type needs support for growth. So staking is necessary. There are many varieties and strains; some are introduced from U.S.A.

Broad Beans (Vicia faba)

The Broad beans are also not very much popular but are grown in some states and hilly regions. The pods are in cluster and it has not branches. The seeds may be yellow, black or white. However it can be grown in plains during September and October.

Sword Beans: The sword beans has deep roots, the leaves are pinnate, trifoliate, dark green and long. They are grown in home gardens. Some are shrubs and some are climbing type. The pods are thick, long, and leathery and look like a sword. The seeds are white or reddish. Tender seeds are used as food and mature ones are cattle feed. The white seeded bushy type is suitable for home gardens.

Goa Bean or Ridge Bean

In case of the Goa bean or winged bean or ridge bean the pods are used as vegetable and each pod has four feathery wings on four sides. It is a creeping type and is mostly grown in home garden particularly in south India and northeast India. The seeds are small with brownish in colour. The dry seeds have a tough seed coat. The pods are green and about 12 to 18 cms long. It is grown in pits and needs organic manure and fertilizers like other beans. To trail the branches it need a Bower or trellis. It grows vigorously and flowering starts within three months. It is an excellent nutritious and delicious vegetable.

Soya Bean (Glucine max)

Soya bean is very popular in Japan and China. It is a versatile bean with high protein content. Its cultivation is now done in India on a wide scale because of its industrial uses. It is grown in sub-tropical regions and in hilly slopes. However, it can be grown in plains during Rabi season. Seeds for planting should be inoculated. It is grown as Kharif crops in some parts. In hills it is sown during April-May. It requires 4-5 months to mature pods.

Bulbs and Tubers

In horticulture, the word bulb includes underground modified stems which are used for propagation e.g. bulb, corm, tuber and rhizome. Plants with tuberous roots are also grouped as bulbous plants. A large number of these plants producing attractive flowers are grown in the hills and are commercially important plants in floriculture. Many types, flower well both in plains and hills but the season of growth and flowering may vary. A bulbous plant has normally three phases during a year-growth, flowering and dormancy. Vegetative growth may precede flowering as in Gladiolus or flowering starts before the leaf emerges, e.g. Amaryllis, Haemanthus. After growth and flowering, the plants in most cases, enter into rest period and the duration of dormancy varies with the type of plants and environmental conditions like temperature and humidity.

Cultivation

Bulbs, corms, tubers and rhizomes are used for vegetative propagation. Various methods and structures used for propagation are as follows.

- Division of clump-Rhizomatous plants the Canna, Alpinia can be divided into small clumps.
- Offsets produced laterally are separated for multiplication.
- Cormlets or bulblets produce new plants.
- Scales of Lilium are separated and used for propagation.
- Pieces of bulb disc with leaves or rhizome also develop new plan
- Bulbil arising from the axils of leaves can produce new plants.

Bulbous plants are grown in beds, shrubbery, herbaceous border, grassy land and in pots and bowl. Bulbs, corms, rhizome or tuber are usually planted when they have shown signs of sprouting after dormancy.

The planting materials are normally placed deep in the soil and the soil around it is gently pressed. The depth of planting varies with the type or plants and size of bulbs. For most of the bulbous plants grown in tropical garden, the planting depth is between 3-10 cm.

Most of these plants sunny situation, while Eucharis, Zephyranthes thrive better in semishade; few types, e.g. Pancratium, Haemanthus, Zephyranthes grow and flower well in sun and semishade. Bulbs, corms and tubers should not come in direct contact with fresh organic manure, rhizomes are not usually affected. Time of planting depends on the season of flowering, environmental condition of the region and condition of the planting material. In the plains, Gladiolus is planted from September to November, whereas in the hills, the planting season is normally from April-June. Amaryllis, Hippeastrum, Haemanthus are planted in January-February to obtain flowers in March-April. Caladium and Canna are put in the ground in May.

After the above ground portion dries out the bulbs are lifted out carefully. The root and shoot are cut and the bulbs are cleaned before storage. They should be stored in dark, dry and airy place until the time of next planting. Storage condition is very important for obtaining healthy bulbs. If the humidity is very high, fungus develops on the bulbs and causes rotting. In a very dry atmosphere the bulbs shrived and may lose viability.

It is not necessary to lift all types of bulbous plants every year. Amaryllis, tuberose, Caladium, Pancratium, Zephyranthes may remain in the ground for three years before they are lifted. But water should be withheld when the bulbs are lying in dormant condition in ground or pot. Gladiolus, on the other hand, has to be lifted out every year after drying of leaves.

Some Common Bulbous Plants

1. *Amaryllis (Amaryllidaceae):* Amaryllis belladona is the only species of the genus, a native of South Africa. It is a plant with strap-shaped, glossy, green leaves and producing few large funnel-shaped flowers on a stout stalk, from March-May. Leaves appear at the end of the flowering season and die out before the winter sets in.

In warm humid region the plants retain leaves throughout the year and bear flower during the season. Offsets, developed from the original bulb are separated for multiplication. Bulbs are planted 5-8 cm deep in December-January and the plants grow well in rich and porous soil. Amaryllis is not disturbed for several years at one place and the plants continue to produce new bulbs and flowers. It is very suitable for planting in border, shrubbery and in pot. Although rose or rosy pink colour is common, white to purplish colour are also found.

2. *Caladium (Araceae):* The genus comprises of about 16 herbaceous perennial species with tuberous rhizome. They are a native of tropical America and are widely grown in warm humid climate for the beautiful and attractive foliage produced during the rainy season. Some of the important species are Caladium bicolor, C.humboldtii, C.picturatum, and C.schomburgkii. Numerous varieties have been raised particularly form C.bicolor.

   Leaves are usually peltate-segittate, stalks variegated, blades are with very many shades of colour, e.g. variegated green, blue green, dark green, light green, spotted with white, red, transparent white, etc. leaf veins may be red, silvery or green. Leaf margins are coloured with purple, white, yellow or red. In some species, shape of the blade is lanceolate-sagittate. Flowers are unisexual.

   The pot compost for planting Caladium should be prepared by mixing loamy soil with leafmould and well-decomposed cow dug manure. It prefers moist but porous soil. Application of liquid manure at frequent intervals enhances growth and improves colour of foliage. Caladium is propagated through division of the tuberous rhizome and rarely by seeds. It is planted in summer and leaves show a fine display of colour in the rainy season.

   The leaves start fading in autumn and water is gradually withheld until they have withered. In dormant condition, the rhizome may be allowed to remain in the soil or dug out and stored in a cool and dry place.

3. *Cooperia (Amaryllidaceae):* The genus is named after Joseph Cooper, an English gardener, comprises about 6

species native of North America, and differing from Zephyranthes by the long perianthtube and erect anthers. They are tender bulbous plants with the habit of Zephyranthes but blooms only at night. Important species are C.drummondii and C.pedunculata. Flowers are fragrant, white, and sometimes tined with red or pink.

Flowers are solitary, the perianth subtended by a bract like spathe. Leaves are long, narrow flat and twisted and appear along with the flower. It is cultivated in a semishady location and in sandy loam soil. Addition of well-rotted compost, sand and charcoal dust in beds or pots is beneficial. Cooperia is propagated by bulbs. It is planted during spring in rock garden, border or in pots and starts blooming during May-June.

4. *Cooperanthes (Amaryllidaceae):* This genus is a product of intergeneric hybrid between species of Cooperia and Zephyranthes, first raised in 1900 by Percy Lancaster at the Agri-Horicultural Society, Alipore, Calcutta. Hybrid between Cooperia oberwettii x Zephyranthes robusta is known as Alipore Beauty, is probably the best known Cooperanthes. Flowers are light lilac, rose or white.

It grows well in well-drained sandy loam soil, rich in organic matter and in semishade. Cooperanthes are propagated by bulbs. They are planted in the spring and the plants start flowering by the end of summer. They are grown both in bed or in pot and are good as cut flowers.

1. *Crinum (Amaryllidaceae):* The genus Crinum, comprises more than 100 species of large and showy flowering bulbous plants. They are closely allied to Amaryllis and distinguished by the longer perianth tube. The species cross freely and many fine hybrids of Crinum, between Crinum and Amaryllis and Crinum and Hymenocallis are known. Some of he important cultivated species and hybrids are C. longifolium, C. mooriei, C. powellii, C. variabile, C. asiaticum, C. augustum, C. careyanum, C. yuccaeflorum, C. giganteum, C. zeylanicum. Flowers are usually white or in shades of red and purple.

   The stems arise from the tunicated bulbs with a more or less elongated neck. Leaves are large, about 150 cm long and 12-15 cm wide, evergreen or deciduous, depending on the species. Flowers are regular, often highly scented, tube narrow, with six segments and usually funnel-shaped. In some species flowers are 30 cm long and 15 cm wide.

Most of the species prefer shade or semishady location for planting. Crinums usually have large bulbs, sometimes as much as 60 to 80 cm long with numerous fleshy roots. If planted in beds, the soil should be dug to a depth of two to three feet, mixed up well with sufficient quantity of rotten cow dung and compost. It can also be grown in large pots containing soil rich in organic matter.

Bulbs should be of good size and planting is done to a depth below the ground level, twice the size of the bulbs. The plant will grow and flower for years, if watered in the summer months and top dressed with fresh loamy soil around new vegetative growth. Bulbs are planted in April to get flower during the rainy season.

2. *Eucharis Grandiflora (E.amazonica) (Amaryllidaceae):* Eucharis is an important bulbous plant, popularly known as `Amazon Lily'. It prefers semi shade and flowers better in plains than on the hills. The bulbs are globular in shape, leaves large lanceolate. The flowers are about 7 cm across, white, sweet, scented, 5-7 blooms appear on a stalk. Eucharis grows better in pots than in the ground and the compost should be rather rich and heavy instead of sandy, but must not be often disturbed. It flowers in the summer and rains and the leaves begin to wither before the winter.

3. *Gladiolus (Iridiaceae):* Gladioli are among the most beautiful flowers, blossoming from October to March in plains and during June to September in the Hills. This genus comprises more than 150 species of perennial herbs with base of stem swollen into a corm. Most of them have their origin in South Africa, although a few originated from Europe. The species in cultivation and of considerable importance include: G. cardinalis, G. childsii, G. colvillei, G. gandavensis, G. lemoinei, G. nanceianus, G. nanus, G. primulinus, G. psittacinus, G. purpureoauratus, G. saundersii. The numerous crosses have evolved large

number of varieties of varying characters and it is almost impossible to keep up with and enumerate the different varieties of gladioli. Wide range of attractive colour shades may be grouped into white, yellow, cream shaded pink, orange, pale pink on white, pink on cream apricot, salmon, cherry red, orange-scarlet, crimson, purple, mauve and violet, etc.

Gladioli prefer sunny situation with light sandy soil, with pH between 6 to 7. If the soil is heavy, addition of river sand and charcoal improves the soil condition. They grow well both in pot and in beds and the magnificent spikes brighten the garden and room as cut flower. Two parts sandy loam soil mixed with one part of each well-rotted cow dung manure and leafmould and a handful of bone meal is recommended for pot culture. If planted in beds the ground should be dug deep and well-decayed compost at the rate of 4 kg per meter mixed thoroughly and left for sun drying at least for a fortnight. At planting time fresh manure should not be used, but a dressing of bonemeal, superphosphate and wood ash may be given with advantage. In the plants, planting is usually done during September-November and in the hills from April-June, but the flowering season may be extended by early and late planting.

Gladioli are propagated by seeds, corms and cormlets. New varieties are raised from seeds. Seeds germinate freely and the seedlings grown carefully, will flower in the second year. Large flower spikes, however, develop after 3-4 years when the corms attain a good size. Gladioli corms are planted 8 to 10 cm deep into the soil, at a distance of 20-30 cm between the rows and the plants. Application of liquid manure, once at the vegetative stage and again after the formation of flower buds has been found very effective.

The flowering spikes appear in 60-90 days after planting, the flowers continue to open in succession from below upwards and the open flowers remain fresh for a number of days. After flowering, the leaves begin to turn yellow and wither. The plants are then lifted with the corms and cormlets and kept in a dry shady place for a week for drying. The corms are thoroughly cleaned, the cormlets separated and stored on a layer of sand in a dry, cool, airy and shady place until the next planting season. The corms should be examined regularly and those showing sign of rotting or fungus growth should be removed. They are also stored in cold storage but in a dry atmosphere.

4. *Gloriosa (Liliaceae):* As the name implies, the genus Gloriose means `full of glory', popularly known as climbing or creeping lily. They are tall creeping plants, support themselves by means of tendril which arises from leaves. This genus comprises about six species of rhizomatous plants native of Africa and tropical Asia. Important species, commonly grown in the gardens in India are G. superba and G. rothschildiana. Leaves are oblong, lanceolate. Flowers are showy on long pedicels in leaf axils, perianth of 6 distinct long segments; stamens six with versatile anthers.

   Well-drained soil in a sunny location is ideal for planting Gloriosa and the attractive flowers make a fine display of colour when trained on bamboo frame work or low trellis. Gloriosa is vegetatively propagated from rhizome, which may be cut in pieces and planted to a depth of 3-4 cm in April-May.

   The plant flowers in July-August. After flowering is over the plant begins to wither and rhizome becomes dormant. If left undisturbed. Gloriosa continues to produce flowers for several years in the same place. Application of liquid manure once at the active phase of vegetative growth and another just before flowering is recommended.

5. *Haemanthus (Amaryllidaceae):* Haemanthus is one of the popular and attractive bulbous plants, commonly known as Football Lily or Bloody Lily. The genus comprises nearly 60 species, native of South and tropical Africa. The flowers are red, crimson, scarlet, pink and some species are white or pale green in colour. Some of the important species are H. multiflorus, H. magnifica, H. lindenii, H. albo-maculaius, H. coccineus, H. tigrinus, and H. candidus.

   Bulbs are usually large with thick skin. Stems green,

short, thick and fleshy. Leaves are usually large and luxuriant, turn yellow and dry in the winter months, the scape is sometimes curiously coloured. Inflorescence is a dense, many flowered umbel, perianth straight and erect with a short cylindrical tube. Flowers are showy and produced in ball-like heads. Fruits berry like, indehiscent. Flowers often appear before the leaves or sometimes simultaneously.

To plant in beds, the soil is deep dug and mixed with well-rotten compost and planted 40 cm apart. It grows well in pot and small pot is preferred. Application of organic manure in the rains and liquid manure before flowering is beneficial and the plants will continue to flower for several years in the same place.

They are propagated by offsets, which should be detached from the mother plant during spring. The bulbs are planted in pots or in beds during spring season to get bloom in summer and early part of rainy season.

6. *Hedychium (Zingiberaceae):* Hedychium consists of 40 species of rhizomatous herbs native of Asia and several species are grown for their beautiful and fragrant flowers in both plains and hills. It prefers semi-shade and moist soil and produces many flowered large spikes during July-October. The leaves die before winter.
7. *Hemerocallis (Liliaceae):* Hemerocallis is popularly known as Day Lily as the flowers last for a day and the blossoms fade at night. This genus consists of more than a dozen species and they are mostly native of China and Japan. The plants are very hardy, stout rooted, glabrous, perennial herbaceous plants, admired for their showy blooms.

   The leaves are almost grass like, 2 ranked at the base of the scape. Flowers are lily like, large funnel shaped, yellow or reddish orange or brown in colour. Though the individual flower last for a day, but many flowers open successively to keep the lasting beauty for a long period.

   Hemerocallis grows in wide variety of soil. While planting, the ground should be dug to a depth of 50 to 60 cm and mixed up with well-rotten compost. They take a year or two to establish properly and should not be disturbed very frequently. Planting is done in February to March and flowering continues from May to August.

Propagation is by division of the clumps and also by seeds. Some of the species are self-sterile and seeds can be obtained easily by crossing two species. Many new-varieties with attractive large flowers of various colours have developed by hybridisation. After 4-5 years, Hemerocallis clumps may be forked out during January-February separated and can be replanted again in the planting season.

8. *Hippeastrum (Amaryllidaceae):* Hippeastrum, commonly known as Royal Dutch Amaryllis is one of the finest flowering bulbous plants. The flowers of Royal Dutch Amaryllis are larger, open widely, about 15-20 cm across, tube is shorter and the plant develops larger bulb than Amaryllis belladona. The spectacular flowers have various shade of bright colour and wide range of variation exists in varieties developed by hybridisation. Hippeastrum produced by Ludwig are famous throughout the world. A double flowered variety is also available.

   The leaves are broader than Amaryllis and do not dry out in the plains. The bulbs are lifted in November-December, stored in dry cool place for a few weeks and replanted in January-February. Hippeastrums develop flower during March-April and application of liquid organic manure before emergence of flower stalk improves the size and quality of flowers.

9. *Hymenocallis (Amaryllidaceae):* This genus comprises about 40 species of bulbous plants native of South America (one in Africa i.e. H. speciosa) cultivated for beautiful fragrant flowers, consisting of narrow green white petals with large coronas and protruding stamens. Some of the important species are H. macrostephana, H. harrisiana, H. speciosa, H. caribaea, H. calathina, H. rotata, H. amancaes, H. tubiflora. Flowers are white, except H. amancaes which is bright yellow.

   Well-drained soil in sunny location is good for planting Hymenocallis. The bulbs are planted just below the surface of the soil during February to get the bloom in the month of May. The soil should be prepared by mixing 2 parts of loamy soil, one part of each well-rotten cow dung manure and leafmould. Watering should be given freely during the growing season. So long the plants remain healthy, transplanting or shifting from pot to pot is not recommended. Whenever necessary bulbs are dug out during September-October and stored in dry place for planting in February next year. Hymenocallis is propagated by bulbs.

10. *Nelumbo (Nymphaeaceae):* Nelumbo or Nelumbium, popularly known as Lotus is a flower of national importance in India. The genus consists of two species- the one bearing yellow flower and the other produces white or pinkish flower.

    The plant has rhizomatous stem, large peltate leaves standing well above the ground and it spreads rapidly in shallow water or wet soil. Like Nymphaea, Lotus prefers soil rich in organic matter and flowers profusely in summer and rains. Lotus do not grow well in lily pools and so tank is ideal for it. N. lutea bears scented yellow flowers 15-20 cm across N. nucifera is common in India. The flowers are white or pink, 20-25 cm in diameter. Lotus is propagated by rhizome or seed.

11. *Nymphaea (Nymphaeaceae):* The Water Lily is the most popular flowering plant in water garden. The genus comprises of 40 species of aquatic, rhizomatous plants widely grown in tank, lily pool, streams, lakes and in suitable earthenware containers.

    The leaves are oval or round in shape, floating on the water or standing above the water surface. The flowers are very attractive, show various colours: white, yellow, pink, red, blue and shades of colours are found in the varieties developed by hybridisation.

    Water Lily prefers clay soil rich in organic matter and sunny location. The rhizomes are planted in soil in the bed of the tank or lily pool or in baskets or pots filled with rich clay soil and then placed in the tank in March-April. In the gardens in cities and towns water lily is often planted in large earthenware container filled with water and a layer of soil. In several species and varieties do not develop in the winter months and the rhizomes remain dormant. Vegetative growth starts in spring and flowers appear during the summer, rains and early winter.

In some species flowers open in day time while in other blooms open at night and close during the day. The species of Nymphaea commonly grown in water garden are as flowers: N. caerulea produces light blue flowers, 7-10 cm across, day blooming. N. lotus, the night blooming water lily, bearing white flower 12-15 cm across. N. odorata, a white flowering and fragrant water lily. The flowers are 8-15 cm across. N. pubescens, has white flowers, 8-12 cm across. N. stellata, a species commonly found in the warm humid region of India. It produces pale blue flowers about 10-16 cm in diameter. Nymphaea is propagated from rhizomes, bulbils and seeds. Most of the species and varieties thrive and flower well at water depth between 50 to 150 cm.

12. *Polyanthes Tuberosa (Amaryllidaceae) Tuberose:* Tuberose (Polianthes tuberosa), a native of Mexico, is widely grown in the plains of India and blooms profusely during the summer and rains, flaunting its fragrance outdoors and indoors. Most artistic garland, floral ornaments, bouquets and buttonholes are made from these flowers. The long spike of flowers is excellent for table decoration. The flowers remain fresh for days together and bathe the atmosphere with their sweet pleasant fragrance.

    The tuber is bulb-shaped and the plant is commonly classed among the `Bulbs'. The leaves are 70-80 cm long, narrow, linear and radical, bright green in colour. The flowering stalk which emerges from the centre of the cluster of leaves is about 80-120 cm long bearing successively smaller long pointed clasping leaves, uppermost ones are much reduced and bract-like. The flower buds are tubular. Flowers are 5-6 cm long, borne in pairs in an open spike pure wavy white, highly fragrant, tube 2.5 to 3.0 cm long, slightly bent near the base, expanding widely where it meets the oblong obtuse segments.

There are three types of tuberose in cultivation-`single' with one row of corolla segments; `semi-double' bearing flowers with two to three rows of segments and `double' having more than three rows of corolla segments. Though there is no popular named variety in tuberose. `The Pearl' is known to be a variety in the double flowered type. A variety with variegated leaf bearing single flowers is also grown. Single flowered type is more widely cultivated than the other types.

The bulbs remain dormant during the winter months in places where the temperature is low and if early planting is desired, the dormancy can be successfully broken by dipping the bulbs in 4% thiourea solution for one hour. Normally tuberose begins to flower in 90-95 days after planting. It flowers during the summer and rains (April-September) in the plains of eastern part of the country and from May to July on the hills, while in milder climate tuberose flowers well throughout the year.

Bulbs having diameter 2.0 and 2.5 cm show satisfactory growth and flowering. The average life of a flower spike is about 10-15 days in situ, while that of an individual flower varies from 4-6 days. Vase life of a spike varies from 7-10 days, depending on the environment and change of water. The spikes remain fresh for a longer period, if kept in 4% sugar solution.

Tuberose can be successfully grown in pots, beds, borders, shrubberies and rockeries. Propagation is by means of seeds and bulbs. Vegetative propagation is commonly practiced and desirable too, because such plants produce better flowers within a short period after planting.

The land should be thoroughly cultivated until the soil comes to good tilth. A good amount of rotted cow dung or farm yard manure should be incorporated with the soil at least 10-15 days before planting. The bulbs are planted 4-5 cm deep in beds in March-April in the plains and in May-June on the hills. The planting distance between the bulbs is 10-15 cm, while the rows are spaced 25 cm apart. For pot cultivation, 1-2 bulbs are planted in a 20 cm pot.

A fertilizer mixture containing 6 gms of urea, 16 gms each of superphosphate (single) and muriate of potash per square meter has been found to show satisfactory growth and flowering. The above mixture should be applied in two equal doses-the first dose before planting and the second one, 4 weeks after sprouting of the bulb. Commercial growers dig out bulbs 3 years after planting. They are stored in a cool, dry and shady place and planted again in the following spring.

13. *Zephyranthes (Amaryllidaceae):* Zephyranthes has a fanciful meaning i.e. the 'West-wind Flower' commonly known as Zephyr Flower or Fairy Lily. The Genus has about 50 species of bulbous flowering plants, native of warmer region of America. It is related to Habranthus, Pyrolirion and Hippeastrum. Some of the popular species are Z. candida, Z. citrina, Z. grandiflora, Z. rosea, Z. macrosiphon, Z. tubispatha and Z. verecunda. The flowers are with many colour ranges i.e., white, yellow and various shades of pink.

Zephyranthes is a hardy bulbous herb. Leaves are filiform or linear or may be strap shaped. Flowers solitary, peduncle elongated slender, hollow; perianth funnel-shaped, erect or slightly inclined, stamen 6, stigma 3-fid.

It is commonly propagated by separating the bulbs, sometimes through seeds. Bulbs are planted in spring, 30 to 40 mm deep and 2 to 15 cm apart. They flower during summer and rains and make a very colourful display especially in rains. Top dressing of organic manure once a year in a rainy season will promote flowering and the bulbs may be left undisturbed for many years.

Bibliography

Adams H.: *Hot Pepper Improvement*, St. Michael, CARDI, 1997.

Agarwal P. K.: *Improvement of Citrus*, New Delhi, Malhotra Publishing House, 1993.

Alexander M. P. and Ganeshan S.: *Pollen Storage*, New Delhi, Malhotra Publishing House, 1993.

Andrews J.: *The Domesticated Capsicums*, University of Texas Press. 1995.

Andrews L.: *Citrus Production - Orange*, St. Augustine, Trinidad and Tobago, 1990

————: *Production of Minor Crops Sapodilla, Tamarind, Guava and Genip*, Port of Spain, Trinidad and Tobago, 1996.

Arguello D. S.: *Production, Post-harvest Management and Exportation of Tropical Fruits in Costa Rica*, Wageningen, Technical Centre for Agricultural and Rural Cooperation, 1992.

Ashworth S.: *Seed to Seed*, Decorah, Seed Savers Publications, 1991.

Barbeau G.: *Tropical Fruits in Nicaragua*, Managua, Nicaragua Ministerio de Desarrollo Agropecuario, Agraria, 1990.

Batlle I. and Tous J.: *Carob Tree (Ceratonia siliqua L.)*. Rome, International Plant Genetic Resources Institute, 1997.

Berger, P.L.: *Pyramids of Sacrifice: Political Ethics and Social Change*, New York, Basic Books, 1974.

————: *Pyramids of Sacrifice: Political Ethics and Social Change*, New York, Basic Books, 1974.

Brooks, D.: *Water: Local-Level Management*, Ottawa, International Development Research Centre, 2002.

Bunnik J. S. C.: *Fresh Fruits and Vegetable: a Survey on the Netherlands and other Major Markets in the European*

Community, Netherlands, Centre for the Promotion of Imports from Developing Countries, 1990.

Burton W.G.: *The Potato*, Holland, H. Veenman & Zonen N.V., 1966.

Cao Van P.: *An Integrated Approach for the Production and Processing of Minor Fruits in Martinique (FWI)*, Port of Spain, Trinidad and Tobago, 1996.

Cardona M. J. G., Carvajal S. L. B., Salinas D. G. C. and Isaza R. G. B. : *Proceedings of the International Seminar on Plantain Production, Quindio, Colombia,* Quindio, Corporation Colombiana de Investigation Agropecuaria, 1998.

Chadha K. L. and Pareek O. P.: *Advances in Horticulture: Fruit Crops,* New Delhi, Malhotra Publishing House, 1993.

Chambers, R.: *Rural Development: Putting the Last First*, London, Longman, 1983.

Collymore L.: *Fruit Production in Barbados*, Port of Spain, Trinidad and Tobago, 1996.

Coste R.: *Coffee: the Plant and the Product*, London, MacMillan, 1992.

Crucefix D.: *Avocado Variety Selection for Export Development,* Roseau, CARDI, 1996.

Cull Brian & Lindsay Pax: *Fruit Growing in Warm Climates for Commercial Growers & Home Gardeners*, Australia, Reed Books, 1995.

Currah L. and Proctor F. J.: *Onions in Tropical Regions*, Kent, Natural Resources Institute, 1990.

Daniells J.: *Illustrated Guide to the Identification of Banana varieties in the South Pacific*, Canberra, ACIAR, 1995.

Degras L.: *Yam: a Tropical Root Crop*, Wageningen, CTA/ MacMillan, 1993.

Dhatt A. S. and Singh Z.: *Propagation and Rootstocks of Citrus*, New Delhi, Malhotra Publishers, 1993.

Diederichsen A.: *Coriander (Coriandrum sativum L.)*. Rome, International Plant Genetic Resources Institute, 1996.

Doijode S. D.: *Seed Germination in Fruits*, New Delhi, Malhotra Publishers, 1993.

Dudley, E.: *The Critical Villager: Beyond Community Participation*, London, Routledge, 1993.

Featherly H. I.: *Taxonomic Terminology of the Higher Plants*, USA, Iowa State College Press, 1954.

Ferentinos L.: *Proceeding of the Sustainable Taro Culture for the Pacific Conference*, Honolulu, HITAHR, 1993.

Forde S: *Proceedings of CARDI / CTA Workshop on Marketability of Caribbean Minor Fruits*, Port of Spain, Trinidad and Tobago, 1996.

Georges, S.: *The Debt Boomerang: How Third World Debt Harms Us All*, Boulder, Westview Press, 1992.

Glowinski Louis: *The Complete Book of Fruit Growing in Australia*, Australia, Lothian Publishing Company Pty. Ltd., 1991.

Godden G.: *Growing Citrus Trees*, Australia, Lothian Publishing Company Pvt. Ltd., 1988.

Gowen S.: *Banana and Plantains*, London, Chapman Hall, 1996.

Green S.K. and Kim J. S.: *Sources of Resistance to Viruses of Pepper (Capsicum spp.): a Catalogue*, Taipei, AVRDC, 1994.

Gunjate R. T. and Tawde A. B.: *Propagation and Rootstocks of Mango*, New Delhi, Malhotra, 1993.

Hartley W.: *A Checklist of Economic Plants in Australia*, Melbourne, C.S.I.R.O., 1979.

Harwood, R. R.: *Marshalling Technology for Development: Proceedings of a Symposium*, Washington, DC, National Academy Press, 1995.

Herklots G. A. C.: *Vegetables in South East Asia*, London, George Allen & Unwin Ltd., 1972.

Hessayon D. G. Dr.: *The Vegetable Expert*, England, PBI, Publications, 1985.

Huaman Z.: *Descriptors for Sweet Potato*, Rome, International Board for Plant Genetic Resources, 1991.

Hurst Jacqui & Rutherford Lyn.: *A Gourmet's Book of Mushrooms & Truffles*, Sydney, Golden Press Pvt. Ltd., 1991.

Jacquat Christiane: *Plants from the Markets of Thailand*, Bangkok, Duang Kamol, 1990.

Jeffers P.: *Evaluation of Four Onion Varieties in Montserrat*, Plymouth, CARDI, 1992.

Kenridge K. C. and Hardy B.: *Biology and Agronomy of Forage Arachis*, Cali, International Centre for Tropical Agriculture, 1994.

Kroll R.: *Cut Flowers*, Wageningen, CTA, 1995.

Kunelius T.: *Annual Ryegrasses in Atlantic Canada*, Ottawa, Agriculture Canada, 1991.

Lofgren, H., Richards, A.: *Food Security, Poverty, and Economic Policy in the Middle East and North Africa*, Washington, DC, International Food Policy Research Institute, 2003.

Mabberley D. J.: *The Plant-Book : a Portable Dictionary of the Vascular Plants*, Cambridge, Cambridge University Press, 1997.

Madulid Domingo A.: *A Pictorial Cyclopedia of Philippine Ornamental Plants*, Philippines, Makati Metro Manila, 1995.

Malins A.: *Postharvest Handling of Pineapple and Mango*, Port of Spain, Trinidad and Tobago, 1992.

Mannetje, L. T. & Jones, R. M.: *Plant Resources of South-East Asia,* Wageningen, Pudoc Scientific Publishers, 1992.

Matsuoka H.: *Cultivation of Panicum Genetic Resources for Evaluation of Characteristics*, Tokyo, JICA, 1997.

Mc Donald F.: *Plant Tissue Culture Manual for Yam, Cassava, Sweet Potato, Dasheen (taro) and Tannia (cocoyam)*, Roseau, Dominica and Cave Hill, 1993.

Mellor, J. W.: *The New Economics of Growth*, Ithaca, Cornell University Press, 1976.

Miller William: *Dictionary of English Plant Names*, London, John Murray, 1884.

Mitra S.: *Postharvest Physiology and Storage of Tropical and Subtropical Fruits*, Oxon, CABI, 1997.

Morris M. L. : *Maize Seed Industries in Developing Countries*, Colorado, Lynne Rienner, 1998.

Morton Julia F.: *Fruits of Warm Climates*, Miami, Julia F. Morton Publisher, 1987.

Murali T. P. and Duncan E. J.: *In Vitro Propagation of Banana through Aseptic Manipulation of Male Inflorescences*, Port of Spain, NIHERST, 1990.

Nabhan G. P.: *Wild Phaseolus Ecogeography in the Sierra Madre Occidental, Mexico: Areographic Techniques for Targeting and Conserving Species Diversity*, Rome, International Board for Plant Genetic Resources, 1990.

Nijdam J. & De Jong A.: *Elsevier's Dictionary of Horticulture in Nine Languages*, Elsevier Scientific Publishing Co. Amsterdam; New York.

Oka H. I.: *Origin of Cultivated Rice*, Elsevier, Japan Scientific Societies Press, 1988.

Oldham P.: *Cost of Production of Major Tree Crops in Dominica*, Roseau, Ministry of Agriculture, 1991.

Ou Yang Jue Ya: *Comparative Study of Mandarin and Cantonese*, Beijing, China Social Sciences Publishing House, 1993.

Percival John: *The Wheat Plant - A Monograph*, London, Duckworth & Co., 1921.

Pilgrim R.: *Post Harvest Handling of Minor Exotics*, St. George's, Grenada, 1996.

Ragone D.: *Breadfruit: Artocarpus Altilis (Parkinson) Fosberg*, Rome, International Plant Genetic Resources Institute, 1997.

Rehm Sigmund: *Multilingual Dictionary of Agronomic Plants*, Boston, Kluwer Academic Publishers, 1994.

Roy S. K.: *Research on Multipurpose Tree Species in Asia: In Vitro Clonal Propagation of Artocarpus Heterophyllus*, Bangkok, Winrock International Institute for Agricultural Development, 1991.

Singh H. P. and Chadha K. L.: *Genetic Resources of Citrus*, New Delhi, Malhotra Publishing House, 1993.

Sinha G. C., Reddy Y. T. N. and Singh G.: *Propagation and Rootstocks in Guava*, New Delhi, Malhotra Publishers, 1993.

Sperling L. and Berkowitz P. *Partners in selection: bean breeders and women bean experts in Rwanda*. Washington, USA: CGIAR, 1994.

Sreekumar V., Indrasenan G and Mammen G.: *Studies of the Quantitative and Qualitative Attributes of Ginger Cultivars*, Calicut, Kasaragod, 1990.

Stover R. H. and Simmonds N. W.: *Bananas*, United Kingdom: Longman Scientific and Technical, 1991.

Thavarasook C.: *Proceedings of the Mungbean Meeting 90, Chiang Mai, Thailand,* Bangkok, Tropical Agricultural Research Centre, 1991.

Thomas E.: *Fruit Production in St. Kitts and Nevis*, Port of Spain, IICA, 1996.

Vargas E. M., Macaya G., Baudoin J. P. and Rocha O. J.: *Variation in the Content of Phaseolin in Wild Populations of Lima Beans (Phaseolus lunatus L.) in the Central Valley of Costa Rica*. Plant Genetic Resources Newsletter, 2000.

Webb M.: *Weed Problems and their Control in Rice, Papaya, Citrus and Sugar cane: a Report to the Belize Plant Protection Service*, United Kingdom, Natural Resources Institute, 1993.

Whealy K.: *The Garden Seed Inventory*, Decorah, Seed Saver Publications, 1988.

Whitwell A.: *Dominica Orchard Crop Management and Research Project: Pest and Disease Management,* United Kingdom, Natural Resources Institute, 1991.

Woolfe Jennifer A.: *The Potato in the Human Diet*, Cambridge, Cambridge University Press, 1989.

Yoshida T.: *Cultivation of Citrus Genetic Resources for Evaluation of Characteristics*, Tokyo, JICA, 1996.

Index

D

E

F

G

H

N

O

P

R

S

T

U

V

W

Y

❑❑❑